HRW

ADVANCED
ALGEBRA

TEACHING RESOURCES

explore

communicate

APPLY

$g(\theta)=15\sin(200\theta)$

$h(t)=-16t^2+27t+240$

HOLT, RINEHART AND WINSTON
Harcourt Brace & Company
Austin • New York • Orlando • Atlanta • San Francisco • Boston • Dallas • Toronto • London

TO THE TEACHER

HRW Advanced Algebra Teaching Resources contains blackline masters that complement regular classroom use of *HRW Advanced Algebra*. They are especially helpful in accommodating students of varying interests, learning styles, and ability levels. The blackline masters are conveniently packaged in four separate booklets organized by chapter content. Each master is referenced to the related lesson and is cross-referenced in the *Teacher's Edition*.

- **Practice Masters** (one per lesson) provide additional practice of the skills and concepts taught in each lesson.
- **Enrichment Masters** (one per lesson) provide stimulating problems, projects, games, and puzzles that extend and/or enrich the lesson material.
- **Technology Masters** (one per lesson) provide computer and calculator activities that offer additional practice and/or alternative technology to that provided in *HRW Advanced Algebra*.
- **Lesson Activity Masters** (one per lesson) connect mathematics to other disciplines, provide family involvement, and address "hot topics" in mathematics education.
- **Chapter Assessment** (one multiple-choice test per chapter and one free response test per chapter)
- **Mid-Chapter Assessment** (one per chapter)
- **Assessing Prior Knowledge and Quiz** (One Assessing Prior Knowledge per lesson and one Quiz per lesson)
- **Alternative Assessment** (two per chapter) is available in two forms, one which entails concepts found in the first half of the chapter and the other which entails concepts found in the second half of the chapter.

Developmental assistance by B&B Communications West, Inc.

Printed in the United States of America

ISBN 0-03-095392-8

2   3   4   5   6   7   066   99   98   97

TABLE OF CONTENTS

Practice & Apply
7.1 Exploring Population Growth

According to the Environmental Protection Agency (EPA), the air quality standard for the pollutant ozone is 0.12 parts per million. The Philadelphia metropolitan area released an average of 8.40 parts per million of ozone in 1990. The area is required to decrease emissions at a projected constant rate of 35% per year.

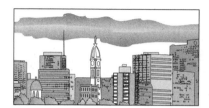

1. Complete the table by finding the projected ozone level for each successive year until the year 2000. Round the ozone level to the nearest hundredth.

Year x	1990	1991	1992	1993	1994	1995
Ozone Level y						
Year x	1996	1997	1998	1999	2000	
Ozone Level y						

2. According to the table, what is the projected ozone level for 1998? _____

3. When will the ozone level meet the air quality standard of 0.12 parts

   per million? _____

4. In each calculator step, what number is multiplied by each year's projected ozone level to obtain the next year's projected ozone level? _____

The number of cable television subscribers in the United States on January 1, 1970 was 2500 and was growing at a rate of 7% per year.

5. What is the growth rate? _____

6. What is the multiplier for this growth rate? _____

7. Find, to the nearest whole number, the projected population of cable

   television subscribers after 20 years. _____

8. Find the function describing the growth of the number of cable television subscribers.

   _____

The retail trade sales totaled $957 billion in the United States in 1980 and was growing at a rate of 7.5% per year.

9. Find the function that models the growth of retail trade sales. _____

10. To the nearest billion, what is the projected total sales for the year 2000? _____

Practice & Apply
7.2 The Exponential Function

1. Arrange $3^{1.5}$, $2^{\frac{1}{4}}$, $4^{\sqrt{2}}$ from smallest to largest. _____

Use a calculator to evaluate each expression. Round to the nearest thousandth.

2. $5^{-2.03}$ _____

3. $11.2^{1.2}$ _____

4. $8^{\sqrt{7}}$ _____

5. $2^{3.1}$ _____

6. Using a graphics calculator, compare the graphs of $f(x) = 4^x$ and $g(x) = 3 \cdot 4^x$.

_____

7. How could you use the graph of $f(x) = b^x$ to sketch the graph of $g(x) = a \cdot b^x$?

_____

A union contract guarantees the members a constant salary increase of $3\frac{3}{4}\%$ per year for 5 years. Jane's initial salary is $18,500.

8. What exponential function models Jane's salary increase? _____

9. To the nearest dollar, what will Jane's salary be in 3 years? _____

Suppose $2500 is put in a savings account that pays an annual percentage rate of 2.75%. Find the balance after each of the following time periods t, and compounding periods n.

10. 1 year; annually _____

11. 5 years; annually _____

12. 1 year; quarterly _____

13. 5 years; quarterly _____

14. 1 year; monthly _____

15. 5 years; daily _____

16. The number of bacteria in a colony is found to quadruple every hour. If there are initially 500 bacteria present, in about how many hours will there be 1 billion bacteria?

_____

Practice & Apply
7.3 Logarithmic Functions

Write an equivalent logarithmic equation.

1. $25^{\frac{1}{2}} = 5$ _____

2. $10^3 = 1000$ _____

3. $(0.25)^2 = 0.0625$ _____

4. $2^{-2} = \frac{1}{4}$ _____

Write equivalent exponential equations.

5. $\log_2 32 = 5$ _____

6. $\log 0.001 = -3$ _____

7. $0 = \log_5 1$ _____

8. $2 = \log_4 16$ _____

9. Which is the inverse of $f(x) = 3^x$? _____

    a. $f^{-1}(x) = \log_3 x$ **b.** $f^{-1}(x) = \frac{1}{3^x}$ **c.** $f^{-1}(x) = 3\log x$ **d.** $f^{-1}(x) = x^3$

Solve each equation and check your answers.

10. $\log_8 (x + 1) = \log_8 (2x - 2)$ _____

11. $\log_3 (3x - 4) = \log_3 (-5x + 8)$ _____

12. $\log_7(6y + 4) = \log_7(-3y - 5)$ _____

13. $\log_{10}\left(\frac{6c + 3}{11}\right) = \log_{10}\left(\frac{3c}{5}\right)$ _____

14. $\log_2\left(\frac{3x - 2}{5}\right) = \log_2\left(\frac{8}{x}\right)$ _____

15. $\log_5(3y^2 - 4y) = \log_5(5y)$ _____

16. Graph $y = \left(\frac{1}{2}\right)^x$ on the grid provided, then sketch y^{-1} on the same grid.

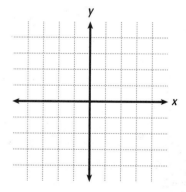

In astronomy, the observed brightness of stars is called apparent magnitude. Two stars can be compared using the formula $d = 2.5\log r$, where d is the difference of the apparent magnitude of the two stars and r is their brightness ratio. The brightest star Sirius A, has an apparent magnitude of -1.46. Alpha Centauri has an apparent magnitude of -2.9.

17. Find the difference between the apparent magnitude of Sirius A and Alpha Centauri. _____

18. Find the brightness ratio of the two stars to the nearest tenth. _____

Practice & Apply
7.4 Exploring Properties of Logarithmic Functions

Express the following as a sum or difference of logarithms.

1. $\log_{10}\left(\frac{1}{3}\right)$ _____

2. $\log_{10} 24$ _____

3. $\log_5 2a^4bc^3$ _____

4. $\log_2\left(\frac{(y-2)^2}{y+1}\right)$ _____

Find the exact value of each expression.

5. $\frac{1}{2}\log_3 81$ _____

6. $5\log_8 64$ _____

7. $\frac{1}{8}\log_a a^8$ _____

Express each as a single logarithm.

8. $\log_{10} 2 - \log_{10} 6$ _____

9. $3\log_{10} 3 + \log_{10} 2$ _____

10. $\frac{1}{2}\log_{10} 25 + \log_{10} 5$ _____

11. $2(\log_{10} 2 - \log_{10} 3 - \log_{10} 1)$ _____

Find the given logarithm without using the log function on your calculator. Assume that $\log_{10} 2 \approx 0.301$ and $\log_{10} 3 \approx 0.477$.

12. $\log_{10} 6$ _____

13. $\log_{10} 81$ _____

14. $\log_{10} \sqrt{3}$ _____

15. $\log_{10}\left(\frac{2}{3}\right)$ _____

Solve for x.

16. $\log_6 x = \log_6 7 + \log_6 6$ _____

17. $\log_2 x - \log_2 3 = \log_2 4$ _____

18. $\log_{10} x + \log_{10}(2x - 3) = \log_{10} 2$ _____

19. $3\log_6 t - \log_6 t = \log_6 2$ _____

The interest rates paid by taxable money market funds averaged 13% in 1988. Since 1988, the interest rate m, has declined according to the assumed function

$$m(x) = \frac{\log_{10} x - \log_{10} 13}{\log_{10} 0.78}, \text{ where } x \text{ is the number of years after 1988.}$$

20. Complete the table by finding the projected interest rates for each successive year until 1998. Round the interest rate to the nearest hundredth.

Number of years after 1988	1	2	3	4	5	6	7	8	9	10
Interest rate (%)										

Practice & Apply
7.5 Common Logarithms

Use your calculator to evaluate each logarithm. Round to the nearest hundredth.

1. log 3000 _____

2. log 3 _____

3. log 30 _____

4. log 300 _____

5. log 25 _____

6. 5log 100 _____

7. log 1 _____

8. log 0.25 _____

Use your calculator to find the number whose common log is the given value. Round to the nearest hundredth.

9. 1 _____

10. −2 _____

11. 0.05 _____

12. 12 _____

13. −1.35 _____

14. 5.07 _____

15. 0.023 _____

16. −0.301 _____

Use a graphics calculator to graph $f(x) = \log(x - 3)$.

17. What are the *x*- and *y*-intercepts? _____

18. Find the domain and range of *f*. _____

19. How does the graph of *f* compare to that of $y = \log x$? _____

Retail sales of cellular and other radio telephones can be modeled by the equation $y = \dfrac{\log x - \log 3542.7}{\log 1.25}$, where *y* is the year with 1989 = 1 and *x* is the sales in log millions of dollars.

20. In what year did sales reach $4334 million? _____

21. In what year did sales reach $8700 million? _____

22. What is the projected year when sales will reach $200,000 million? _____

The Richter scale is a system for rating the severity of an earthquake. The Richter energy number R of an earthquake is defined as $R = 0.671\log(0.37E) + 1.46$, where *E* is the energy in kilowatts per hour released by the earthquake.

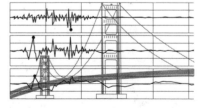

23. If the energy released by an earthquake is 16,124,385 kilowatts per hour, what is the Richter energy number to the nearest tenth? _____

24. The strongest recorded earthquake in American history was 80 miles east of Anchorage, Alaska in 1964. The energy released was approximately 61,600,000,000 kilowatts per hour. What is the Richter energy number to the nearest tenth? _____

Practice & Apply
7.6 The Natural Number, e

Evaluate each expression to the nearest hundredth.

1. $\ln 100$ _____

2. $\ln e^{(-3)}$ _____

3. $e^{0.1}$ _____

4. $e^{\ln 1}$ _____

Write equivalent logarithmic or exponential equations.

5. $e^0 = 1$ _____

6. $0.37 = e^{(-1)}$ _____

7. $\ln 14.73 = 2.69$ _____

8. $\ln 0.5 = -0.69$ _____

Solve each equation for x without graphing.

9. $2.7^x = 10$ _____

10. $2^{3x} = 25$ _____

11. $5^{-x} = 0.3$ _____

The population P of a town is growing according to the function
$P = 10,000\, e^{\frac{x}{30}}$, where x is the number of years.

12. About how many people will live in the town in 10 years? _____

13. In about how many years will there be 35,000 people in the town? _____

14. In approximately how many years will the population double? _____

The average typing speed per minute is found by dividing the
total words typed by the time. A business school advertises
that the average typing speed S, will improve according to the
function

$$S = \frac{120}{1 + 5.4e^{(-0.12x)}}$$ after x weeks of instruction.

15. Find the average typing speed after 5 weeks of instruction. _____

16. Find the average typing speed after 12 weeks of instruction. _____

Four months after it stopped advertising, a manufacturing company
noticed that its sales per unit y, had dropped each month according to the
function $y = 100,000e^{(-0.05x)}$, where x is the number of months after the
company stopped advertising.

17. Find the projected drop in sales per unit six months after the
company stops advertising.

_____

18. What appears to happen to the sales per unit decline as the
number of weeks after advertising ceased increases?

_____

Practice and Apply
7.7 Solving Exponential and Logarithmic Equations

Solve the following equations.

1. $2^x = 2^5$ _____

2. $\log_5 25 = x$ _____

3. $\log_3(x - 3) = 2$ _____

4. $\ln (x - 2) = \ln 4$ _____

5. $2\ln x = 0$ _____

6. $\ln 3x = 3$ _____

The net profit of a leading computer company is modeled by $p(x) = -7.84 \ln (x) + 3.35$, where x is the number of years after 1990 and $p(x)$ is the net profit in billions of dollars.

7. What is the projected net profit for 1995? _____

8. In what year did the computer company lose approximately 5 billion

dollars? _____

Use a calculator to solve each equation to the nearest hundredth.

9. $3^x = 10.4$ _____

10. $\log x^{-1} = 0.04$ _____

11. $e^{3x} = 31.8$ _____

12. $\ln x^{\frac{1}{3}} = 2$ _____

13. $5^{2x + 1} = 98$ _____

14. $\ln 3x = 0.1$ _____

The voltage drop in tests simulating nonstop use of a motorized toy for heavy-duty batteries is modeled by the equation $V(x) = 1.3(0.77)^x$, where x is the number of hours and $V(x)$ is the voltage. If the voltage drops below 0.9 volts, the heavy-duty battery ceases to power electrical devices.

15. Find the voltage after 1 hour of nonstop use.

_____

16. Find the approximate number of hours of nonstop use before the voltage drops below 0.9 volts.

_____

Enrichment
7.1 Exploring Population "Ungrowth"

You have seen how populations can grow at a given rate. As this example shows, population can also decline in size at a given rate.

Suppose a population of a certain type of bacteria is present in your bloodstream when you see your doctor. Your doctor starts you on antibiotics, which kill off 2% of the bacteria present each hour. What percent of the original bacteria population is left after 24 hours?

Set up a table using the fact that if 2% of the population is destroyed, then 98% remains at the end of each hour.

Time Past (hours)	Amount Remaining
0	100%
1	1.00(0.98) = 98%
2	0.98(0.98) = 0.9604 ≈ 96%
3	0.9604(0.98) = 0.941192 ≈ 94%
4	0.941192(0.98) = 0.92236816 ≈ 92%
5	0.92236816(0.98) = 0.9039207968 ≈ 90%

Continuing this pattern, after 24 hours there will be about 62% of the original bacterial population remaining.

Use a calculator to solve each of the following population problems.

1. An urban city has a population of 2,000,000. The population is expected to decline at an annual rate of 3.8% because people are moving to the suburbs from the city. What will the city's population be in 20 years?

   _____

2. A farmer discovers a new insect has invaded her soybean crop. She consults with the local farm bureau and is advised to spray with a certain chemical that will destroy the insect population at a rate of 5.6% per hour. About what percent of the insect population will remain after 12 hours?

   _____

3. A chemist has 10 grams of a radioactive substance. If 50% of the substance decays (loses its radioactivity) every 6 days, how many grams of radioactive substance will remain after 180 days?

   _____

4. A small town has a population of 250 people. If 7.3% of the residents leave each year, in how many years will the population of the town be less than 50?

   _____

5. Suppose a college department has 750 graduates one year, but after that, the number graduating from that department declines by 3.7%. In how many years will there be less than 500 graduates in that department?

   _____

Enrichment
7.2 Present Value

Given a specific interest rate and compounding schedule, suppose you know how much money you want to have several years from now. The amount you should deposit today to meet your goal is known as *present value*.

For example, if your bank compounds interests semiannually at 6% per year, you can use the compound interest formula to determine how much you should deposit today in order to have at least $5000 in four years.

In the formula $A(t) = P\left(1 + \dfrac{r}{n}\right)^{nt}$, notice that the principal P, is the present value you are looking for.

$5000 \le P\left(1 + \dfrac{0.06}{2}\right)^{2(4)}$ Substitute the given quantities.

$5000 \le P(1.03)^8$ Simplify.

$\dfrac{5000}{1.03^8} \le P$ Use a calculator.

$P \ge 3947.05$

You should deposit $3947.05 in order to have at least $5000 in four years.

Find the present value for each situation given the desired amount at the end of each time period.

1. $4000; 5% interest is compounded quarterly for 4 years. _____

2. $5000; 6% interest is compounded monthly for 5 years. _____

3. $8000; 4% interest is compounded quarterly for 10 years. _____

4. $500; 6% interest is compounded monthly for 18 years. _____

5. $15,000; 6% interest is compounded semiannually for 21 years. _____

6. $20,000; $4\frac{1}{2}$% interest is compounded annually for 21 years. _____

7. $12,000; 8% interest is compounded annually for $8\frac{1}{2}$ years. _____

8. $25,000; 6% interest is compounded quarterly for 25 years. _____

9. $100,000; 4% interest is compounded annually for 65 years. _____

Enrichment

7.3 Solving a New Type of Equation

You can use the following fact to solve certain equations involving logarithms.

$$\boxed{\text{If } \log_c a = \log_c b, \text{ then } a = b.}$$

For example, if $\log_{10}(2x + 8) = \log_{10}(4x - 3)$ then

$$2x + 8 = 4x - 3$$
$$2x + 11 = 4x$$
$$11 = 2x$$
$$x = 5.5$$

Remember that the value or values of x must make the logarithmic expression positive.

Solve each equation for x.

1. $\log_{10}(x^2 - 2x + 3) = \log_{10}(x^2 + 5x - 8)$

2. $\log_{10}(2x - 3) = \log_{10}(x^2 + x - 4)$

3. $\log_{10}(2x + 7) = \log_{10}(5x - 9)$

4. $\log_{10}(5x - 9) = \log_{10}(x^2 + 2x - 7)$

5. $\log_9(4x - 7) = \log_9(x^2 - 9x + 2)$

6. $\log_8(3x - 6) = \log_8(9x - 15)$

7. $\log_{10}(2x - 2) = \log_{10}(x^2 - x - 14)$

8. $\log_{10}(7x - 4) = \log_{10}(x^2 - 12x + 1)$

9. $\log_3(2x + 1) = \log_3(x^2 - 2x - 4)$

10. $\log_4(7x - 1) = \log_4(x^2 + 9)$

11. $\log_{10}(x^2 + x - 1) = \log_{10}(x^2 - 9x + 5)$

12. $\log_{10}(2x + 9) = \log_{10}(17)$

Enrichment
7.4 More Equations Involving Logarithms

You can use the properties of logarithms and the fact that if $\log_c a = \log_c b$, then $a = b$ to solve logarithmic equations.

For example, to solve the equation $\log_6 4 + \log_6 x = 48$, first simplify using the properties of logarithms.

$$\log_b x + \log_b y = \log_b xy, \text{ and if } \log_c a = \log_c b, \text{ then } a = b.$$

$$\log_6 4 + \log_6 x = \log_6(4x)$$
$$\log_6(4x) = \log_6 48$$
$$4x = 48$$
$$x = 12$$

Solve each equation for x.

1. $2\log_8 3 - \dfrac{1}{2}\log_8 16 = \log_8 x$

2. $\log_3(x - 1) + \log_3(x - 2) = 1$

3. $\log_4 16 - \log_4 2 = \log_4 x$

4. $\dfrac{1}{2}\log_{10} 25 = \log_{10} x$

5. $\log_8 4 + \log_8 x = \log_8 24$

6. $\dfrac{1}{3}\log_5 x = 1$

7. $\dfrac{1}{2}\log_8 64 = \log_8 x$

8. $\log_6 27 - \log_6 9 = \log_6 x$

9. $\log_4 x + \log_4 2x = \log_4 32$

10. $\log_{10} x + \log_{10} 4 + \log_{10} 2x = \log_{10} 64$

11. $\log_{10} 6 + \log_{10}(x - 3) = \log_{10} 3x$

12. $\log_{10} x = \dfrac{1}{3}\log_{10} 8 - \dfrac{1}{2}\log_{10} 144$

Enrichment
7.5 The Parts of a Common Logarithm

1. Complete the table. Write each number in scientific notation; next use a calculator to find each common logarithm to three decimal places; then express each logarithm as the sum of a whole number and decimal.

Number	Scientific Notation	Common Log	Log Expressed as Sum
3	3×10^0	0.477	0 + 0.477
30			
300			
3000			
300,000			

2. What do you notice about the decimal part of each logarithm?

_____

3. How is the whole number part of each logarithm related to the number written in scientific notation?

_____

The whole number part of a logarithm is called the characteristic. The decimal part of the logarithm is called the mantissa.

$$\underset{\underset{\text{characteristic}}{\uparrow \quad \uparrow \qquad \uparrow}}{\log 3 = \log (3 \times 10^0) = 0 + 0.477} = \overset{\overset{\text{mantissa}}{\downarrow}}{0.477}$$

4. Given that log 2 = 0.301 and log 7 = 0.845, complete the table by filling in the characteristic and mantissa of each number.

Number	Characteristic	Mantissa
20,000		
7,000,000		
0.2		
0.07		
14		
1.4		

Enrichment
7.6 Approximating *e*

Over the years, mathematicians have discovered a number of ways to approximate the value of *e*. One way is to evaluate the expression $1 + \frac{1}{1!} + \frac{1}{2!} + \ldots + \frac{1}{x!}$. For example, to evaluate for $x = 4$, you would find the value of $1 + \frac{1}{1!} + \frac{1}{2!} + \frac{1}{3!} + \frac{1}{4!}$.

Evaluate the expression for each given value of *x*.

$$\left(1 + \frac{1}{x}\right)^x$$

1. $x = 1$ _____   **2.** $x = 2$ _____   **3.** $x = 3$ _____

4. $x = 4$ _____   **5.** $x = 5$ _____   **6.** $x = 6$ _____

7. $x = 7$ _____   **8.** $x = 8$ _____   **9.** $x = 9$ _____

10. $x = 10$ _____   **11.** $x = 100$ _____   **12.** $x = 1000$ _____

$$1 + \frac{1}{1!} + \frac{1}{2!} + \ldots + \frac{1}{x!}$$

13. $x = 1$ _____   **14.** $x = 2$ _____   **15.** $x = 3$ _____

16. $x = 4$ _____   **17.** $x = 5$ _____   **18.** $x = 6$ _____

19. $x = 7$ _____   **20.** $x = 8$ _____   **21.** $x = 9$ _____

22. $x = 10$ _____   **23.** $x = 11$ _____   **24.** $x = 12$ _____

25. Which expression gives a better approximate value of *e*?

_____

26. Which expression is easier to evaluate on a calculator for large values of *x*? Why?

_____

27. What procedure or method simplifies the calculations in the factorial expression approximating *e*?

_____

_____

Enrichment

7.7 Depreciation

Businesses own many items that depreciate, or go down in value, over time— buildings, machinery, cars, and so on. Depreciation plays an important part in figuring out how much profit a business has made.

For example, a company may buy a building for $50,000. If the building depreciates at a rate of 5% per year, it might be useful to know when the building will be worth $20,000.

Using the formula $A = P(1 + r)^n$:

$$20,000 = 50,000(1 - 0.05)^n \qquad \text{Use a minus sign to show depreciation.}$$
$$0.4 = 0.95^n \qquad \text{Simplify.}$$
$$\log 0.4 = n \log 0.95 \qquad \text{Convert to a logarithmic equation.}$$
$$n = \frac{\log 0.4}{\log 0.95} \qquad \text{Use a calculator.}$$
$$\approx 17.86$$

The building will be worth $20,000 in about 18 years.

Solve each depreciation problem.

1. XYZ Company bought a machine that costs $150,000. If the machine depreciates at a rate of 8% per year, in about how many years will the machine be worth $40,000?   _____

2. Megan's Muffins bought a new oven for $950. Four years later, the oven was worth $200. What was the annual rate of depreciation to the nearest tenth of a percent?   _____

3. Jason uses his car for his job. He is allowed to depreciate the car 8% per year. If the car was worth $23,500 new, in about how many years will the car be worth $3000?   _____

4. Adam's Construction bought a new piece of equipment for $245,000. If the equipment depreciates at a rate of 12% per year, about how much will it be worth in four years?   _____

5. Ashley bought a computer for $3000. If the computer depreciates at a rate of 25% per year, about how much will the computer be worth in three years?   _____

6. Farms Unlimited bought a tractor in 1995 for $75,000. If the tractor depreciates at a rate of 15% per year, what year will the tractor be worth about $20,000?   _____

Technology
7.1 An Effect of Small Changes in the Base

A spreadsheet is particularly useful if you want to compare b^x and c^x for a given value or values of x. One question that you might ask is how different b^x and c^x are when c and b are close to one another. To answer the question, set up a spreadsheet that looks like the one shown. Column D will contain the formula for the difference of the corresponding values in columns B and C. For example, D2 will contain B2−C2.

	A	B	C	D
1	X	B^X	C^X	DIFFERENCE
2	1			
3	2			
4	3			
5	...			

Use a spreadsheet and x = 1, 2, 3, ..., 20.

1. Let b = 1.1 and c = 1.0. Find the value in column D for x = 1. _____

2. Let b = 1.1 and c = 1.0. Find the value in column D for x = 20. _____

3. Let b = 10.1 and c = 10.0. Find the value in column D for x = 1. _____

4. Let b = 10.1 and c = 10.0. Find the value in column D for x = 20. _____

5. Let b = 100.1 and c = 100.0. Find the value in column D for x = 1. _____

6. Let b = 100.1 and c = 100.0. Find the value in column D for x = 20. _____

7. Suppose that $c = b - 0.1$. How does the difference in column D change as x increases?

    _____

8. Suppose that $c = b - 0.1$. What effect does the 0.1 difference have if b is small? If b is large?

    _____

Technology
7.2 Rational Exponents

In Lesson 7.2 you learned that if $b > 0$, then b^x is an exponential expression that has meaning for any real number x. This means that b^x is defined whenever x is an integer. It also means, however, that b^x is defined when x is any rational number as well.

What exactly does b^x mean when $x = 0.5$ or 1.5, for example? To answer the question, you can use a spreadsheet or graphics calculator. The spreadsheet shown contains a series of nonnegative rational numbers in column A. Column B contains the formula 3^X.

You may be surprised to learn that if x is rational and b is positive, b^x is a number that you can find by using powers and roots.

	A	B
1	X	3^X
2	0.00	
3	0.25	
4	0.50	
5	0.75	
6	1.00	
7	1.25	
8	1.50	
9	...	

Create a spreadsheet like the one shown extending column A and B until $x = 5.0$. In Exercises 1–6, evaluate each expression with a graphics calculator. Then give the cell name whose entry matches your calculator answer, if there is one.

1. $\boxed{\sqrt{}}\,\boxed{(}\,3\,\boxed{\wedge}\,1\,\boxed{)}$

2. $\boxed{\sqrt{}}\,\boxed{(}\,3\,\boxed{\wedge}\,4\,\boxed{)}$

3. $\boxed{\sqrt{}}\,\boxed{(}\,3\,\boxed{\wedge}\,2\,\boxed{)}$

_____   _____   _____

4. $3\,\boxed{\wedge}\,(7/3)$

5. $3\,\boxed{\wedge}\,(6/3)$

6. $3\,\boxed{\wedge}\,(3/3)$

_____   _____   _____

7. Based on your results from Exercises 1–6, describe how to find $3^{\frac{n}{2}}$, where n is a positive integer.

_____

8. Describe how to use powers and roots to find $3^{\frac{11}{2}}$ and $3^{\frac{11}{3}}$.

_____

9. Describe how to use powers and roots to find $3^{\frac{7}{5}}$. How would you use a calculator to evaluate the expression?

_____

Technology
7.3 Logarithms and Powers of 10

In Lesson 7.3, you learned that every positive number has a logarithm whose base is 10. You can use a spreadsheet to find the logarithm of a positive number x. In many spreadsheets, you can use the formula LOG(X). If, for example, you enter 14.5 in column A and LOG(14.5) in the same row but in column B, the entry in column B will be 1.161368. If, however, you enter a series of positive numbers into column A and compute their logarithms in column B, you will see patterns emerge.

Set up a spreadsheet that contains a series of positive numbers in column A and their logarithms in column B. In Exercises 1–8, find the logarithm of each number using base 10 logarithms.

1. 1.6   **2.** 18.2   **3.** 132   **4.** 1.32

_____   _____   _____   _____

5. 1820   **6.** 16   **7.** 1320   **8.** 1600

_____   _____   _____   _____

9. Complete the following: If x is a times a positive power of 10, 10^n,

then the logarithm in base 10 of x is related to that of a by _____

In Exercises 10–17, find the logarithm of each number using base 10 logarithms.

10. 452,000   **11.** 379   **12.** 37.9   **13.** 93

_____   _____   _____   _____

14. 379,000   **15.** 4.52   **16.** 0.93   **17.** 452

_____   _____   _____   _____

18. Complete the following: If x is a divided by a positive power of 10,

10^n, then the logarithm in base 10 of x is related to that of a by _____

Technology
7.4 Graphics Calculators and Properties of Logarithms

There are many ways to see that $\log (ab) = \log a + \log b$, where a and b are positive numbers. You may be surprised to learn that you can use a graphics calculator to construct a convincing argument that shows the equation above is true.

Although you cannot graph $y = \log(ab)$ and $y = \log a + \log b$, where y, a, and b are all variables on a graphics calculator, you can graph functions like $y = \log (3x)$ and $y = \log x + \log 3$, $y = \log (4.3x)$ and $y = \log x + \log 4.3$, and so forth.

Use a graphics calculator to graph each pair of functions on the same display screen. Use XMIN = 0 and XMAX = 9.4.

1. $y = \log(3x)$ and $y = \log x + \log 3$

2. $y = \log (4.3x)$ and $y = \log x + \log 4.3$

3. $y = \log (6x)$ and $y = \log x + \log 6$

4. $y = \log (3.9x)$ and $y = \log x + \log 3.9$

5. How are the graphs in each pair related? What does your observation suggest about the two expressions for y?

_____

Use a graphics calculator to graph each pair of functions on the same display screen. Use XMIN = 0 and XMAX = 9.4.

6. $y = \log \left(\dfrac{x}{3}\right)$ and $y = \log x - \log 3$

7. $y = \log \left(\dfrac{x}{6.1}\right)$ and $y = \log x - \log 6.1$

8. $y = \log \left(\dfrac{6}{x}\right)$ and $y = -\log x + \log 6$

9. $y = \log \left(\dfrac{3.9}{x}\right)$ and $y = \log 3.9 - \log x$

10. How are the graphs in each pair related? What does your observation suggest about the two expressions for y?

_____

Technology
7.5 Linear Approximations to Common Logarithms

If you graph $y = \log x$ over the interval $1 \leq x \leq 10$, you will see that the graph is a curve. Thus, the common logarithm function is not linear.

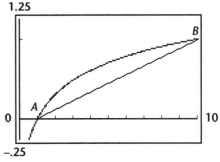

With a spreadsheet, you can explore the question of how much different the logarithmic graph is from a straight line. The graphics calculator display shows the graph of $y = \log x$ along with the straight line segment connecting $A(1, 0)$ and $B(10, 1)$.

1. Write an expression for the y-coordinate of a point on the line containing A and B.

   _____

2. Create a spreadsheet in which column A contains $x = 1, 2, 3, ..., 9$ and 10, column B contains LOG(X), column C contains the values of your expression from Exercise 1, and column D contains the differences of the values in columns B and C.

3. Describe the trend in the values in column D of your spreadsheet. Do you think the y-coordinates of points on $\overline{AB}$ give good approximations to $\log x$? Explain.

   _____

Suppose that point C has coordinates $C(5, 0.699)$.

4. Write an expression for the y-coordinate of a point on the line

   containing A and C. _____

5. Write an expression for the y-coordinate of a point on the line

   containing C and B. _____

6. Modify your spreadsheet in Exercise 2 so that column C contains the values of your expression from Exercise 4 for $1 \leq x \leq 5$, and Exercise 5 for $5 \leq x \leq 10$. Are the approximations to $\log x$ improved? Explain.

7. If you place $A = P_1, P_2, P_3, ... P_{n-1}$, and $P_n = B$ on the graph of $y = \log x$ and join them in order, would you expect the y-coordinates of these segments to give you better approximations to $\log x$ than the y-coordinates of $\overline{AC}$ and $\overline{CB}$? Explain.

   _____

Technology
7.6 Different Ways to Calculate *e*

In this activity, you can explore different ways to use mathematics and technology to approximate *e*.

1. Create a spreadsheet like the one shown. Record the sum you get when you fill twelve rows.

	A	B	C
1		1	
2	1		
3	2		
4	3		
5	4		
6	...		

Cell A3 contains 1+A2.
Cell B2 contains B1/A2.
Then use the FILL DOWN command to continue columns A and B.
Cell C1 contains the sum of the entries in column B.

_____

2. Create a spreadsheet with $n = 1, 2, 3, ..., 20$ in column *A* and the value of $\left(1 + \frac{1}{n}\right)^n$ in column B. What number do the entries in column B seem to get closer and closer to?

_____

3. Use a spreadsheet or calculator to find the value of the continued fraction shown through several divisions.

_____

$$1 + \cfrac{2}{1 + \cfrac{1}{6 + \cfrac{1}{10 + \cfrac{1}{14 + ...}}}}$$

4. Use a scientific calculator to evaluate $1 + \frac{1}{1!} + \frac{1}{2!} + \frac{1}{3!} + \frac{1}{4!} + \frac{1}{5!} + \frac{1}{6!}$.

What number does the sum suggest? (The expression *n*! means the product of the integers from 1 to *n*.)

_____

5. Use a scientific calculator to evaluate the expression:

$$1 + \frac{3}{1!} + \frac{9}{2!} + \frac{27}{3!} + \frac{81}{4!} + \frac{243}{5!} + \frac{2187}{6!}.$$

What power of *e* does the sum approximate?

_____

Technology
7.7 Exponential Functions and Linear Equations

The diagram shows a graphics calculator display obtained by graphing $y = 2^x$ and $y = 1.2x + 2$. The display models the exponential-linear equation $2^x = 1.2x + 2$. It is clear from the display that the equation has two solutions. By using the trace feature, you will find that $x = -1.3$ and $x = 2.2$.

You can experiment with other exponential-linear equations to find out whether such an equation always has two solutions.

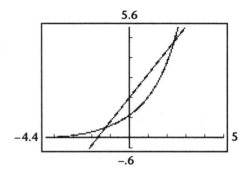

Use a graphics calculator to solve each equation.

1. $2^x = 1.2x + 2.5$

2. $2^x = -2x - 1$

3. $2^x = 2$

_____

_____

_____

4. $2^x = -3$

5. $2^x = 0.6x - 2$

6. $2^x = -1.6x + 3$

_____

_____

_____

7. $2^x = -3x + 2$

8. $2^x = 1.2x + 0.8$

9. $2^x = 3.55$

_____

_____

_____

10. How many solutions can $2^x = mx + b$ have if $m > 0$, $m = 0$, and $m < 0$.

_____

Use a graphics calculator to solve each equation.

11. $0.8^x = -2x + 4$

12. $0.8^x = 2.2x + 2$

13. $0.8^x = -3$

_____

_____

_____

14. $0.8^x = 1.2x - 2$

15. $0.8^x = 2$

16. $0.8^x = -0.4x + 1$

_____

_____

_____

17. How many solutions can $0.8^x = mx + b$ have if $m > 0$, $m = 0$, and $m < 0$.

_____

Lesson Activity
7.1 Residual Plots

When trying to fit a curve or a line to a data set, the better the mathematical model, the less it misses any of the data points. The amount by which a model misses the data points used to generate the model is known as a residual. Examination of residuals helps determine how well a model fits the data points.

Use a graphics calculator to examine the residuals in the bacterial population growth from Exploration 1 in your text.

1. Using the linear regression feature, what is the linear function that best fits the data points?

    _____

2. Let y_p be the y-coordinate of that linear regression function. Find the residuals by completing this table.

Hour (x)	0	1	2	3	4	5	6
Population (y)	100	200	400	800	1600	3200	6400
Population (y_p)							
Residual ($y - y_p$)							

3. How do the residuals compare to the actual y-values in size? _____

4. Which residuals, if any, stand out? _____

5. What trend or pattern, if any, do the residuals seem to follow as the x-values vary?

    _____

6. Using the exponential regression feature, what exponential function best fits the data points? (Round coefficients to the nearest whole number.)

    _____

7. Describe the residuals. _____

8. Compare the two types of models. Which type of regression model had the smallest residuals?

    _____

9. What type of function best models the bacterial population growth?

    _____

Lesson Activity
7.2 Modeling Taxed Savings

In a taxed savings program, you pay income tax each year on the interest or dividends earned on the investment.

Suppose Gail invests $1000 in a certificate of deposit at a bank on her 21st birthday. The one-year certificate pays 5% interest compounded annually, and her income-tax rate is 15%. If Gail plans to invest this amount for 40 years, how can you determine how much will be available after taxes have accumulated on her investment?

Start by making a table to show the growth of the certificate fund and the after-tax accumulation at the end of each year.

Years	Gail's age	Balance (in dollars)	Total Amount (in dollars)
0	21	1000	1000.00
1	22	$1000(1 + 0.05) - 1000(0.05)(0.15) =$ $1000[(1 + 0.05) - (0.05)(0.15)] =$ $1000[1 + 0.05(1 - 0.15)] =$	1042.50
2	23	$1042.50(1 + 0.05) - 1042.50(0.05)(0.15) =$ $1042.50[1 + 0.05(1 - 0.15)] =$ $1000[1 + 0.05(1 - 0.15)][1 + 0.05(1 - 0.15)] =$ $1000[1 + 0.05(1 - 0.15)]^2 =$	1086.81
$\cdot$ $\cdot$ $\cdot$	$\cdot$ $\cdot$ $\cdot$	$\cdot$ $\cdot$ $\cdot$	
n	$n + 21$	$1000[1 - 0.05(1 - 0.15)]^n$	

Notice that the exponent indicates the number of times the interest after taxes has been compounded. Thus, the accumulated balance after 40 years would be $1000[1 + 0.05(1 - 0.15)]^{40} = \$5284.97$.

1. Find the after-tax accumulation after 40 years if the interest rate is compounded quarterly. _____

2. Find the after-tax accumulation after 40 years if the interest rate is compounded daily. _____

3. What is the effect of a change in the number of times the interest rate is compounded? _____

4. Suppose Gail's income tax rate is 18%. What is the after-tax accumulation after 40 years, if the interest is compounded annually at 5%? _____

5. What is the effect of the increased tax rate? _____

Lesson Activity
7.3 Curve Fitting

Jordi has just purchased a new car for $13,250. He plans to keep it for 5
years and then trade it in. To get an approximation of what the trade-in
value will be at the end of 5 years, Jordi consults the *Blue Book* and makes a
table showing the depreciation for prior years.

Age (in years) (x)	0	1	2	3	4	5
Value (in $) ($y$)	13,250	11,527.50	10,928.93	8725.16	7590.90	6604.08

Jordi plots the data points using a graphics calculator. Then he uses the
linear regression features and finds that the linear function that best fits
these data points has a correlation coefficient of −0.991928. Using the
exponential regression feature, the correlation coefficient is −0.991503.

1. Find the linear regression model for Jordi's data. (Round coefficients to
the nearest whole number.) Then find the differences between the
actual y-values and the y-values using the linear regression model.

_____

2. Find the exponential regression model for the data. (Round coefficients
to the nearest whole number.) Then find the differences between the
actual y-values and the y-values using the exponential regression
model.

_____

3. Which type of regression model has the smaller differences? _____

To find a best-fit model, many statisticians draw a scatter plot of the
ordered pairs (x, log y). If the diagram shows a linear relationship, the
original data may be approximated by an exponential function.

4. Find log y for each of the values in
Jordi's table.

5. Graph the ordered pairs (x, log y) on
the grid provided. Does the graph
appear to be linear?

_____

6. Use the linear regression feature to
model the new data points. What is
the correlation coefficient?

_____

7. What best-fit curve would you choose?

_____

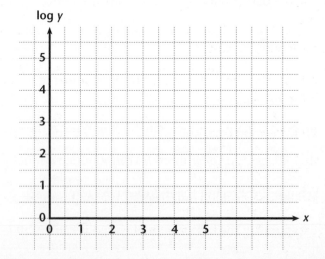

Lesson Activity
7.4 Logarithmic Patterns

The log $I = -0.245\log d + 3$ relates light intensity to the depth below the water's surface.

1. Complete the table using the equation $\log I = -0.245\log d + 3$.

log d	0	0.3	0.6	0.9	1.2	1.5
log I	3	2.9265				

2. Graph the relationship between light intensity and depth on the grid provided.

3. Describe the graph.

_____

4. What is the slope and the y-intercept of this function?

_____

If the graph is nearly linear, then values for d and I may be approximated by the power function, $y = 10^b x^m$, where b is the y-intercept and m is slope of the linear regression function for light intensity.

5. Find the power function for the data set in Exercise 1.

_____

Examine the data points for the relationship between light intensity and distance.

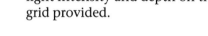

Distance in ft (x)	0.5	1	1.5	2	2.5	3
Intensity (y)	2.4	0.6	0.267	0.15	0.096	0.067

6. Find the values of log x and log y for the data in the table. _____

7. Plot the points (log x, log y) on a separate sheet of graph paper and

describe the graph. _____

8. Find the slope and y-intercept of the graph. _____

9. Explain why the best-fit model is a power function. _____

10. Use the power regression feature to find the function. _____

Lesson Activity
7.5 Modeling pH

A logarithmic scale is a scale in which the units are spaced so that the ratio between successive units is the same. The figure shows three examples of logarithmic scales.

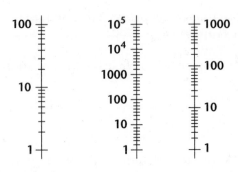

Logarithmic scales are often used to model data with a wide range of values. The pH scale is a logarithmic scale in which the hydrogen ion concentration, $[H^+]$, is written as a power of 10.

1. Change the $[H^+]$ concentration to pH using the formula $pH = -\log[H^+]$. Then complete the scale.

pH												
$[H^+]$	10^0	10^{-1}	10^{-2}	10^{-3}	10^{-4}	10^{-5}	10^{-6}	10^{-7}	10^{-8}	10^{-9}	10^{-10}	10^{-11}

2. Acid rain is an environmental issue of great concern. Find out the meaning of acid rain and how it can change the pH level of lakes. Describe your findings.

_____

_____

_____

3. The normal laboratory value of pH in blood is 7.35–7.45. A decreased or increased value can be associated with certain diseases. Research the following medical conditions: dehydration, starvation, intoxication, diabetic acidosis, chronic renal failure, diabetes insipidus. Determine whether the pH level of blood is decreased, increased, or normal for the conditions listed. Describe your findings.

_____

_____

_____

_____

Lesson Activity
7.6 Estimating ln2

When computer programmers write programs that evaluate natural logs, they try to find a method that saves computer time, reduces round-off errors (errors from repeating decimals), and controls truncation errors (errors from deleting the tail end of a series). Some programs and calculators evaluate ln 2 as 0.6931471806 by approximating the sum of an infinite series.

One method uses the series: $\dfrac{1}{2} - \dfrac{1}{3} + \dfrac{1}{4} - \dfrac{1}{5} + \dfrac{1}{6} - \dfrac{1}{7} + \ldots$.

1. Complete the chart.

Series	Partial Sum	Cumulative Sum
$1 - \dfrac{1}{2} + \dfrac{1}{3} - \dfrac{1}{4} + \dfrac{1}{5} - \dfrac{1}{6} +$		
$\dfrac{1}{7} - \dfrac{1}{8} + \dfrac{1}{9} - \dfrac{1}{10} + \dfrac{1}{11} - \dfrac{1}{12} +$		
$\dfrac{1}{13} - \dfrac{1}{14} + \dfrac{1}{15} - \dfrac{1}{16} + \dfrac{1}{17} - \dfrac{1}{18} +$		
$\dfrac{1}{19} - \dfrac{1}{20} + \dfrac{1}{21} - \dfrac{1}{22} + \dfrac{1}{23} - \dfrac{1}{24}$		

2. What happens to the partial sums as the terms decrease in size?

_____

3. What is a disadvantage of this method? _____

Another method uses the series: $\dfrac{1}{2} + \dfrac{1}{(3)(4)} + \dfrac{1}{(5)(6)} + \dfrac{1}{(7)(8)} + \dfrac{1}{(9)(10)} + \ldots$.

4. Complete the chart.

Series	Partial Sum	Cumulative Sum
$\dfrac{1}{2} + \dfrac{1}{(3)(4)} + \dfrac{1}{(5)(6)} + \dfrac{1}{(7)(8)} + \dfrac{1}{(9)(10)} + \dfrac{1}{(11)(12)} +$		
$\dfrac{1}{(13)(14)} + \dfrac{1}{(15)(16)} + \dfrac{1}{(17)(18)} + \dfrac{1}{(19)(20)} + \dfrac{1}{(21)(22)} + \dfrac{1}{(23)(24)} +$		
$\dfrac{1}{(25)(26)} + \dfrac{1}{(27)(28)} + \dfrac{1}{(29)(30)} + \dfrac{1}{(31)(32)} + \dfrac{1}{(33)(34)} + \dfrac{1}{(35)(36)}$		

5. What happens to the partial sums as the terms decrease in size?

_____

6. Do you think this method will approach ln 2 faster or slower than the first method?

_____

Lesson Activity
7.7 "Zooming In" on Exponential and Logarithmic Equations

Some equations involving exponential or logarithmic functions are difficult to solve by ordinary algebraic means. These equations may be solved using a graphical approach.

Use your graphics calculator to graph $2x^2 - x + \ln x = 0$. Make sure all other plots are turned off. Your graph should look similar to the one shown. The display shows that there is one region that might contain a solution.

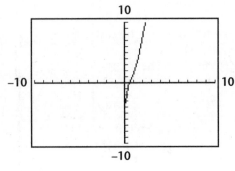

Select ZDecimal from the Zoom menu. Then select Zoom In and position the cursor on the apparent point of intersection. Notice that the x-coordinate has one decimal place. Press ENTER. Then move the cursor until it's over the point of intersection. The x-coordinate should now have one more decimal place.

Use Zoom In once more. The x-coordinates will have more decimal places. By continuing to zoom in you can approximate the root to a greater number of decimal places.

1. Use this method to find all the solutions to the equation $xe^{-x} + 2e^{-x} = 0$ to the nearest tenth.

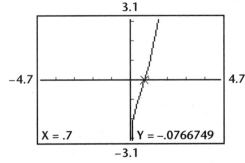

The flow rate y in ft^3/s over a 4-foot wide rectangular dam in a stream is modeled by the equation $y = 12.43x^{1.5}$, where x is the water level above the dam.

2. Use a graphics calculator to graph the function. Then use the trace feature to find the flow rate to the nearest hundredth when the water level is 1.5 ft above the dam.

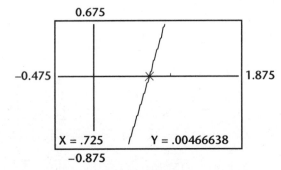

3. According to the graph, what is the water level above the dam when the flow rate is approximately 117.4 ft^3/s?

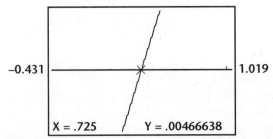

Assessing Prior Knowledge
7.1 Exploring Population Growth

1. Simplify the expression $3x(2 - y)$. _____

2. Use the Distributive Property to factor out the greatest common factor of $8xy + 4x^2 - 16xy^2$.

   _____

- -

Quiz
7.1 Exploring Population Growth

Identify the multiplier for each growth rate.

1. 0.7% _____   2. −5.03% _____   3. 8% _____

Suppose $4000 is invested at 6.25% interest compounded once per year.

4. What is the growth rate for the $4000 invested? _____

5. What is the multiplier for this growth rate? _____

6. What will the value of the original investment be after 8 years?

   _____

7. How long will it take for the original investment to double? _____

8. Write the function that models the growth of the investment.

   _____

9. How many years will it take for $3000 invested at an annual interest of 9% to exceed the $4000 investment at 6.25%, when both investments are compounded once per year.

   _____

Assessing Prior Knowledge
7.2 The Exponential Function

1. Express $a^{0.3}$ using rational exponents. _____

2. Express $a^{\frac{3}{10}}$ using integer exponents. _____

- -

Quiz
7.2 The Exponential Function

Classify each function as polynomial or exponential.

1. $f(x) = x^4$ _____

2. $f(x) = 5^x + 1$ _____

3. $f(x) = 3x - 2$ _____

Use a calculator to evaluate each expression to the nearest hundredth.

4. $3^{\sqrt{5}}$ _____ 5. $2.3^{5.24}$ _____

6. Graph $f(x) = 5^x$ on the grid provided.

7. Explain how the graph of the function $f(x) = 5^x - 2$ can be obtained from the graph of $f(x) = 5^x$.

_____

_____

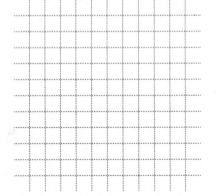

8. If \$2000 is invested at an annual interest of 4.5% compounded monthly, how much will the investment be worth in 10 years?

_____

9. Tell whether the functions $f(x) = \left(\frac{1}{4}\right)^x$ and $g(x) = 4^{-x}$ are equivalent. Explain.

_____

Assessing Prior Knowledge
7.3 Logarithmic Functions

1. Find the inverse relation for $f(x) = x^2$. _____

2. Find the inverse of $f(x) = 2x + 3$. _____

- -

Quiz
7.3 Logarithmic Functions

Write an equivalent logarithmic equation for each exponential equation.

1. $5^{-2} = \frac{1}{25}$ _____

2. $4^3 = 64$ _____

3. $36^{\frac{1}{2}} = 6$ _____

Write an equivalent exponential equation for each logarithmic equation.

4. $2 = \log_4 16$ _____

5. $\log_3\left(\frac{1}{9}\right) = -2$ _____

6. $\log 10{,}000 = 4$ _____

7. Use the grid provided to graph $f(x) = 4^x$ and sketch $f^{-1}(x)$ on the same set of axes.

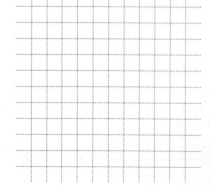

Solve each equation for x.

8. $\log_4(5x - 2) = \log_4(2x + 3)$ _____

9. $\log_{10}(x^2 - 2x) = \log_{10}(3x - 6)$ _____

Assessing Prior Knowledge
7.4 Exploring Properties of Logarithmic Functions

Simplify each expression.

1. $2^3 \times 2^4$ _____

2. $\dfrac{5^8}{5^5}$ _____

3. $((3)^2)^3$ _____

- -

Quiz
7.4 Exploring Properties of Logarithmic Functions

Express each as a sum of logarithms.

1. $\log_4(3 \cdot 5)$ _____

2. $\log_{10} x(x + 5)$ _____

3. $\log_6 xy^3 z^2$ _____

Express each as a difference of logarithms.

4. $\log_2\left(\dfrac{25}{16}\right)$ _____

5. $\log_3\left(\dfrac{vw}{xy}\right)$ _____

6. $\log_{10}\left(\dfrac{x^4}{y^3}\right)$ _____

7. Find the exact value of $3\log_2 8$. _____

Express each as a single logarithm.

8. $\log_3 4 + \log_3 8$ _____

9. $(\log_2 15 - \log_2 5) + \log_2 2$ _____

10. Assume that $\log_{10} 3 \approx 0.477$ and $\log_{10} 7 \approx 0.845$. Evaluate $\log_{10} 21$

without using a calculator. _____

Mid-Chapter Assessment
Chapter 7 (Lessons 7.1–7.4)

Write the letter that best answers the question or completes the statement.

_____ **1.** The multiplier for the growth rate 10.2% is

     a. 0.102   **b.** 10.2   **c.** 1.102   **d.** 0.01102

_____ **2.** The y-intercept of the exponential function $f(x) = 2(4.3)^x$ is

     a. 4.3   **b.** 2   **c.** 1   **d.** 0

_____ **3.** The logarithmic equation equivalent to $4^{-3} = \dfrac{1}{64}$ is

     a. $\log_{-3}\left(\dfrac{1}{64}\right) = 4$ **b.** $\log_4(-3) = \dfrac{1}{64}$

     c. $\log_{\frac{1}{64}}(-3) = 4$ **d.** $\log_4\dfrac{1}{64} = -3$

_____ **4.** What is the value of x, if $\log 18 - \log 3 = \log 2 + \log x$?

     a. 4   **b.** 3   **c.** 13   **d.** 19

5. Assume that $\ln 3 \approx 1.099$, $\ln 4 \approx 1.386$, and $\ln 5 \approx 1.609$. Evaluate $\ln 60$ without using a calculator.

_____

6. Solve $C = 10^{15p}$ for p in terms of C. _____

7. A new car sells for $16,300. Suppose the value of the car depreciates after t years according to the formula $V(t) = \$16,300\left(\dfrac{3}{4}\right)^t$. Determine the value of the car four years after it is purchased.

_____

8. The population growth rate is described as -4.02%. Identify the multiplier for this growth rate and describe what is happening to the population.

_____

_____

Assessing Prior Knowledge
7.5 Common Logarithms

1. $10^0 =$ _____

   $10^1 =$ _____

   $102 =$ _____

2. $10^{-1} =$ _____

   $10^{-2} =$ _____

- -

Quiz
7.5 Common Logarithms

Use a calculator to evaluate each logarithm. Round to the nearest hundredth.

1. $\log 374$ _____

2. $\log 0.245$ _____

3. $\log 3826$ _____

4. $\log \left(\dfrac{2}{9}\right)$ _____

Use a calculator to find the number whose common log is the given value. Round to the nearest hundredth.

5. 2.6 _____

6. 0.021 _____

7. -0.83 _____

Use your graphics calculator to graph $f(x) = \log(x + 3)$.

8. Find the domain and range of f. _____

9. How does the graph of f compare with that of $y = \log x$?

_____

Assessing Prior Knowledge
7.6 The Natural Number, *e*

Irrational numbers are nonterminating and nonrepeating. Is 1.333 ...
rational or irrational?

_____

- -

Quiz
7.6 The Natural Number, *e*

**Use a calculator to evaluate each expression to the nearest
hundredth.**

1. ln 5.27 _____

2. $e^{-2.4}$ _____

3. $e^{\sqrt{3}}$ _____

**Write an equivalent logarithmic or exponential equation for
each equation.**

4. $e^{-0.4} = 0.67032$ _____

5. ln 3 = 1.0986 _____

Solve each equation for *x* without graphing.

6. $0.36^{-x} = 5$ _____

7. $5^{3x} = 36$ _____

Use a graphic calculator to graph $f(x) = \ln\left(\frac{x}{3}\right)$.

8. Find the *x*-intercept of *f*. _____

9. How does the graph of *f* compare with that of $g(x) = \ln x$.

_____

Assessing Prior Knowledge
7.7 Solving Exponential and Logarithmic Equations

Complete.

1. $\log_2 2^4 =$ _____

2. $\log_5 25 =$ _____

3. $2^{\log_2 7.4} =$ _____

4. $e^{\ln 2} =$ _____

- -

Quiz
7.7 Solving Exponential and Logarithmic Equations

Solve each equation without using a calculator.

1. $5^x = 5^3$ _____

2. $\log_5 x = 3$ _____

3. $\ln(5x - 2) = \ln 23$ _____

Use a calculator to solve each exponential equation to the nearest hundredth.

4. $3^x = 45$ _____

4. $e^{3x} = 30$ _____

Use a calculator to solve each logarithmic equation to the nearest tenth.

6. $\log x = 7.8$ _____

7. $\ln 3x = 5$ _____

Use a calculator to solve each equation to the nearest hundredth.

8. $3 \ln(3x - 1) = 2.5$ _____

9. $\ln \sqrt{x} = 4.7$ _____

10. $6^{x-2} = 3^{x+1}$ _____

Chapter Assessment
Chapter 7, Form A, page 1

Write the letter that best answers the question or completes the statement.

_____ 1. The population of a northeastern U. S. city in 1990 was 765,834 and said to be growing at a rate of 1.9% per year. What is the projected population for the year 2000?

    a. 780,384   **b.** 924,435
    c. 632,156   **d.** 7,803,848

_____ 2. If $3000 is invested at an annual interest of 4.5% compounded monthly, how much will the investment be worth in 10 years?

    a. $4,700.98   **b.** $590,304.52   **c.** $4,693.13   **d.** $4,704.81

_____ 3. A logarithmic equation equivalent to $64^{\frac{1}{3}} = 4$ is

    a. $\log_{\frac{1}{3}} 64 = 4$ **b.** $\log_{64} 4 = \frac{1}{3}$

    c. $\log_{64}\left(\frac{1}{3}\right) = 4$ **d.** $\log_4 64 = \frac{1}{3}$

_____ 4. The solution to the equation $\log_6(3x - 2) = \log_6(x + 8)$ is

    a. 1.5   **b.** 3   **c.** 5   **d.** 2.5

_____ 5. The exact value of $3\log_4 64$ is

    a. 1   **b.** 9   **c.** 48   **d.** 6

_____ 6. The expression $\log_3 7 + \log_3 4$ is equivalent to

    a. 28   **b.** 11   **c.** $\log_3 28$ **d.** $\log_3 11$

_____ 7. To the nearest tenth, the number whose common log is 3.5 is

    a. 0.54   **b.** 1.25   **c.** 33.1   **d.** 3162.3

_____ 8. An equation equivalent to $e^{0.3} = 1.3499$ is

    a. $\log 1.3499 = 0.3$ **b.** $\log 0.3 = 1.3499$
    c. $\ln 1.3499 = 0.3$ **d.** $\ln 0.3 = 1.3499$

_____ 9. The solution of $2.5^x = 17$ is

    a. 3.09   **b.** 6.8   **c.** 4   **d.** 0.32

_____ 10. The solution of $4^{3x} = 64$ is

    a. 5.3   **b.** 1   **c.** 3   **d.** $\frac{1}{3}$

Chapter Assessment
Chapter 7, Form A, page 2

_____ **11.** To the nearest hundredth, the solution of $3\left(1 + e^{\frac{x}{2}}\right) = 78$ is

    a. 9.41   **b.** 1.61   **c.** 9.66   **d.** 6.44

_____ **12.** What is the y-intercept for $f(x) = 3e^{-4x}$?

    a. 1   **b.** 3   **c.** 0   **d.** −4

_____ **13.** If $5000 is invested at an annual interest of 6% compounded quarterly, how long will it take the investment to double?

    a. 12 years   **b.** $11\frac{1}{2}$ years   **c.** $11\frac{3}{4}$ years   **d.** 11 years

_____ **14.** Consider the graphs of $f(x) = 3^x$ and $g(x) = 3^x + 1$. Which of the following best describes the relationship between the graphs?

    a. The graphs are unrelated.
    b. The graph of $f(x)$ is 1 unit to the left of the graph of $g(x)$.
    c. The graph of $g(x)$ is 1 unit above the graph of $f(x)$.
    d. The graph of $f(x)$ is 1 unit above the graph of $g(x)$.

_____ **15.** Assume that $\log_{10}5 = 0.6990$ and $\log_{10}10 = 1$. Then $\log_{10}50 =$

    a. 0.6990   **b.** 10.6990   **c.** 1.6990   **d.** 6.9900

_____ **16.** If a van sells for $28,400 when new and depreciates after t years according to the formula $V(t) = 28,400\left(\frac{3}{4}\right)^t$, what is the value of the van 3 years after it was purchased?

    a. $11,981.25   **b.** $21,300.75   **c.** $19,170.42   **d.** 15,975.50

_____ **17.** An exponential equation equivalent to $\ln 3 = 1.10$ is

    a. $10^{1.10} = 3$ **b.** $e^3 = 1.10$ **c.** $e^{1.10} = 3$ **d.** $1.10^3 = 10$

_____ **18.** A certain bacteria colony doubles every 8 hours. If the initial colony contains 40 bacteria, how many are present after 5 days?

    a. 16.075
    b. 1,310,720
    c. 12,000,360
    d. 1,320,360

Chapter Assessment
Chapter 7, Form B, page 1

The population of a midwestern city was 782,563 in 1990 and declining at a rate of 1.4% per year.

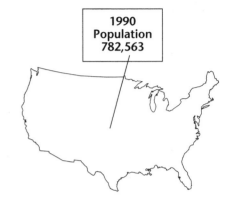

1990
Population
782,563

1. What was the multiplier for the city's population decline?

_____

2. Find the projected population of the city for the year 2000.

_____

3. Use a calculator to evaluate $1.7^{2.68}$ to the nearest thousandth.

_____

4. If $3000 is invested at an annual interest of 5.5% compounded quarterly, find the value of the investment after 10 years.

_____

5. Find an equivalent logarithmic equation for $3^{-4} = \frac{1}{81}$.

_____

6. Solve: $\log_5(3x - 2) = \log_5(5x - 6)$ _____

7. Use a calculator to evaluate $\log 0.0027$, rounded to the nearest hundredth.

_____

8. Use a calculator to find the number whose common log is -1.2, rounded

to the nearest hundredth. _____

9. Use a calculator to evaluate $\ln 15$, rounded to the nearest hundredth.

_____

Chapter Assessment
Chapter 7, Form B, page 2

10. Write a logarithmic equation, equivalent to $e^{0.75} = 2.12$.

_____

11. Solve: $5^{x+1} = \dfrac{1}{625}$ _____

12. Use your calculator to solve $6\left(2 + e^{\frac{x}{4}}\right) = 75$ to the nearest hundredth. _____

13. A radioactive element decays according to the decay function $A = A_0 e^{(-0.05t)}$, where A_0 is the amount of substance initially present and t is the time in years from the present. How long will it take a sample of the element to decay to $\dfrac{1}{4}$ of its original amount?

_____

14. A fully loaded car sells for $42,500 when new and depreciates after t years according to the formula $V(t) = 42,500\left(\dfrac{7}{8}\right)^t$. What is the value of the car 5 years after it was purchased?

_____

15. Assume that $\log_{10} 8 = 0.9031$ and $\log_{10} 6 = 0.7782$. Without using your calculator, find the value of $\log_{10} 48$.

_____

16. Consider the graphs of $f(x) = 5^x$ and $g(x) = 5^x - 1$. Describe the relationship between the graphs of $f(x)$ and $g(x)$.

_____

17. Write an exponential equation equivalent to $\log 28 = 1.4472$.

_____

18. A type of bacteria triples every 48 hours. If the initial colony contains 60 bacteria, how many bacteria are present after 12 days?

_____

Alternative Assessment
Exponential and Logarithmic Functions, Chapter 7, Form A

TASK: To identify the behavior of exponential functions and logarithmic functions by inspection and by graphing

HOW YOU WILL BE SCORED: As you work through the task, your teacher will be looking for the following:

- whether you can identify an exponential growth or decay function
- how well you can describe how logarithmic functions and exponential functions are related

1. Graph $f(x) = 125\left(\frac{1}{5}\right)^{x}$.

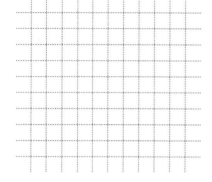

2. What is the domain and range of this function?

_____

3. Explain why f is an exponential function.

_____

_____

4. Describe how you can determine whether f is an exponential growth or

decay function. _____

_____

5. Compare the graph of $f(x) = \left(\frac{2}{3}\right)^{x}$ and $g(x) = \left(\frac{3}{2}\right)^{x}$. Describe how you can

use the graph of f to graph g. _____

_____

6. Graph $f(x) = 3^{-x}$ and $g(x) = \log_{\frac{1}{3}} x$. Explain why the graphs
   are symmetrical with respect to the line $y = x$.

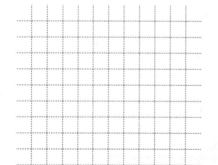

_____

_____

SELF-ASSESSMENT: Compare the graphs of linear, polynomial, and exponential functions. How are they alike? How are they different?

Alternative Assessment
Solving Exponential and Logarithmic Equations, Chapter 7, Form B

TASK: To solve exponential and logarithmic equations

HOW YOU WILL BE SCORED: As you work through the task, your teacher will be looking for the following:

- how well you can solve logarithmic and exponential equations by graphing
- whether you can solve logarithmic and exponential equations by using the exponential-log inverse properties

1. Describe how you would find the solution to $\log(x + 9) - \text{lox } x = 1$ by graphing. Then find the solution.

   _____

   _____

2. Solve $6^{x-1} = 216$. Describe how you can check your solution.

   _____

   _____

Solve each equation.

3. $\ln e^x = 2$

4. $\log x = \frac{1}{3}\log 27$

5. $2^{10-x} = 4^{2+x}$

   _____   _____   _____

6. Solve $\log(3x + 1) - \log(2x + 3) = \log 2$ by graphing. Then solve the equation by using the exponential-log properties. Explain the result.

   _____

   _____

   _____

SELF-ASSESSMENT: For what values is $\log \frac{2x + 4}{3x}$ defined? For what values is the logarithm positive?

Practice & Apply
8.1 Exploring Special Right Triangles

Find x and y.

1.

2.

3.

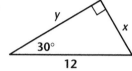

_____ _____ _____

Find the area of each figure.

4.

5.

6.

_____ _____ _____

7. Two guy wires of equal length are attached 50 ft from the base of a tower that is perpendicular to the ground. Each wire makes an angle of 60° with the ground. What is the length of each wire?

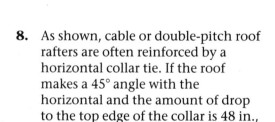

_____

8. As shown, cable or double-pitch roof rafters are often reinforced by a horizontal collar tie. If the roof makes a 45° angle with the horizontal and the amount of drop to the top edge of the collar is 48 in., what is the length of the collar tie?

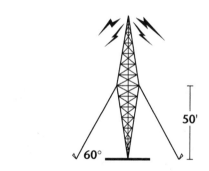

_____

9. Dale wants to use two hooks to hang a 21-inch long picture so that the hook will not show. He places the hooks 3 in. from the vertical side so that the cord makes an angle of 60° with the horizontal. How long is the cord?

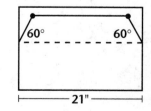

_____

Practice & Apply
8.2 The Unit Circle

Use your calculator to find *x* and *y* to the nearest hundredth.

1.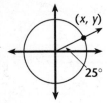

2.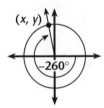

3.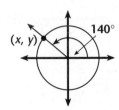

_____   _____   _____

Match each terminal side with an angle in standard position.

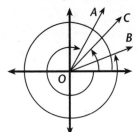

4. 45° _____

5. −300° _____

6. 380° _____

Name the coordinates of each point.

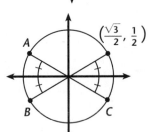

7. *A* _____

8. *B* _____

9. *C* _____

Find the coterminal angle for each angle.

10. 36° _____   **11.** −420° _____   **12.** 300° _____

In physics, vectors are used to describe forces and velocities. For example, vector **OA** describes a velocity of 1 m/s at an angle of 30° south of east.

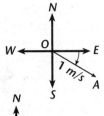

Write the directions for each vector.

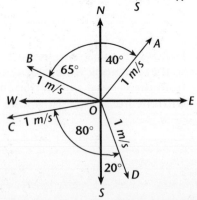

13. vector **OB** _____   **14.** vector **OC** _____

15. vector **OD** _____   **16.** vector **OA** _____

Practice & Apply
8.3 Trigonometric Functions

Find two angles such that $0° \leq \theta < 360°$ for each of the following inverse relations. Round to the nearest degree.

1. $\tan^{-1}(1)$ _____

2. $\sin^{-1}(-\frac{1}{2})$ _____

3. $\cos^{-1}(\frac{1}{2})$ _____

4. $\sin^{-1}(0)$ _____

5. $\cos^{-1}(-0.7)$ _____

6. $\tan^{-1}(1.2)$ _____

7. $\sin^{-1}(0.75)$ _____

8. $\cos^{-1}(-0.2)$ _____

Find the measure of angle A to the nearest degree.

9.

120

97

10.

20

5

11.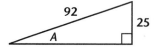

92

25

A

_____   _____   _____

12. A ramp 18 ft long rises to a loading platform that is 3 ft above the ground. Find to the nearest tenth of a degree, the angle that the ramp makes with the ground.

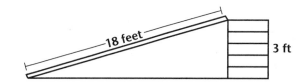

_____

13. Tyne walks 5 m due east and then 12 m due north. To the nearest degree, what direction is Tyne from her starting point?

_____

14. Find to the nearest tenth of a degree, the angle of elevation of the sun, if a tower 150 ft high casts a shadow 210 ft long.

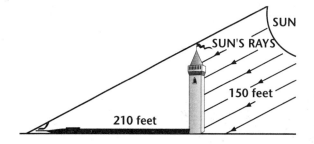

SUN

SUN'S RAYS

150 feet

210 feet

_____

15. The distance from the foot of a hill to its top measured along a straight road is 1000 ft. The height of the hill is 158 ft. Find the angle at which the road rises with the ground to the nearest degree. _____

Practice & Apply
8.4 Exploring Sine and Cosine Graphs

1. The graph shows a model for two waves traveling in the same medium for the same length of time. Explain how the graphs are alike and how they are different.

_____

_____

_____

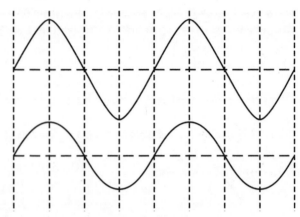

Complete the table by finding the amplitude, period, and phase shift for each function.

2. $y = \cos\left(x - \frac{\pi}{2}\right)$

3. $y = -2\sin 2x$

Amplitude	Period	Phase Shift

The normal monthly mean temperature (in degrees Fahrenheit) for Atlantic City, New Jersey based on a standard 30-year period from 1961–1990 is shown in the table.

Jan	Feb	Mar	Apr	May	June	July	Aug	Sept	Oct	Nov	Dec
30.9	33.0	41.5	50.0	60.4	69.4	74.7	73.4	66.1	54.9	45.8	35.8

4. Use your own graph paper to sketch a scatter plot for the data. Then fit a smooth curve to the data.

5. What is the amplitude of the curve? _____

6. What is the period of the curve? _____

Write a function for each graph.

7.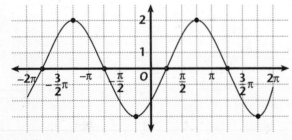

8.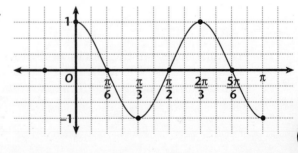

_____   _____

_____   _____

Practice & Apply
8.5 Radian Measure

Find each value to the nearest hundredth using your calculator in radian mode.

1. cos 1.75 _____

2. tan 3 _____

Give the exact value for each of the following.

3. $\cos \frac{7\pi}{4}$ _____

4. $\tan \frac{3\pi}{4}$ _____

Complete the table.

5.

Degrees	Radians	Sine	Tangent
	1.05		
30			
	0.262		
	0.131		
3.75			
1.875			

Identify the vertical shift, amplitude, period (in radians), and phase shift for each of the following functions. Then sketch the graph.

6. $f(x) = 4\cos(x + \pi) + 1$ _____

7. $f(x) = -2 + \sin 3\left(x - \frac{1}{2}\pi\right)$ _____

Write a function for each graph.

8.

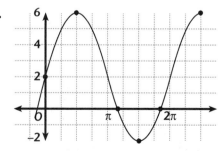

_____

9.

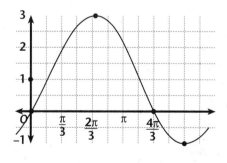

_____

Practice & Apply
8.6 Arc Length and Sector Area

1. Find the length of an arc intercepted by a $\frac{7}{12}$ rotation on a circle of

   radius 4 cm. _____

2. Find the area of the sector of a circle with a central angle that is $\frac{1}{3}$ a

   rotation on a circle with radius of 3 m. _____

3. Find the radius of a circle if the length of the arc intercepted by a $\frac{1}{8}$

   rotation is π m. _____

4. Find the length of $\overset{\frown}{ABC}$ and the area of the shaded region.

   _____

   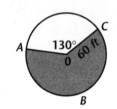

5. The cutting speed of a drill is the distance in inches traveled in 1 minute by the outer corners of the cutting edges of a drill, where D is the diameter of the drill and A is the outer corner of the drill. Find the cutting speed

   in inches per minute of a $\frac{7}{8}$ in. diameter high-speed

   drill that rotates at a rate of 250 rpm.

   _____

   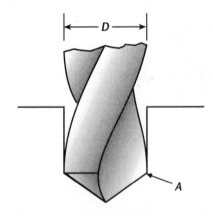

Two pulleys, one 6 in. and the other 24 in. in diameter are connected by a belt. The larger pulley revolves at the rate of 150 rpm and the smaller pulley at the rate of 600 rpm.

6. Find the velocity, in inches per minute, for each pulley at its outer edge that turns the belt.

   _____

7. Find the central angle for a sector with an area of 144 cm^2 in a circle

   with a radius of 13 cm. _____

Practice & Apply
8.7 Applications of Periodic Functions

The yearly change in seasons can be modeled by a sine function as shown. In the diagram, the sky appears flat. The ecliptic, the apparent path of the sun, is shown as a curve that crosses the equator at the two equinoxes.

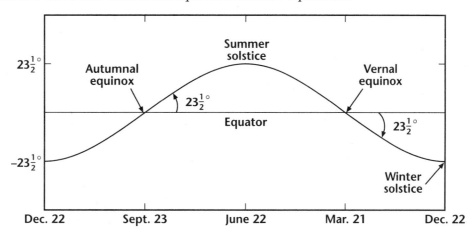

1. Find the domain, range, and period of the function. _____

2. At what angle does the ecliptic intersect the equator? _____

3. When is the sun highest in the sky? _____

4. Describe the path of the sun from the vernal equinox to the winter solstice.

_____

The range of a projectile is modeled by $y = \frac{(2200)^2}{32} \sin 2x$ where 2200 ft/s is the initial velocity, x is the angle of elevation, and y is the range of a projectile.

5. Use a graphics calculator to sketch the graph on the grid provided.

6. Identify the amplitude. _____

7. Give the period. _____

The sales of a seasonal product are modeled by $y = 74 + 40 \sin \frac{\pi}{6}x$, where y is measured in thousandths of units and x is the time in months, with $x = 1$ corresponding to January.

8. Use a graphics calculator to sketch the graph of the given function.

9. Find the amplitude, period, and vertical shift of the function. _____

Enrichment
8.1 Determining Special Angles

Find sin A, cos A, and tan A for each triangle. Use these ratios to determine the measure of angle A.

1.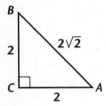

_____

2.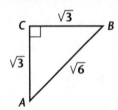

_____

3.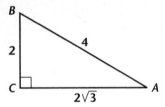

_____

4.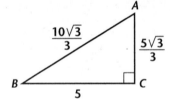

_____

5.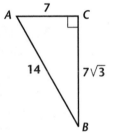

_____

6.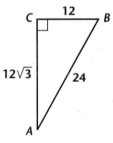

_____

7.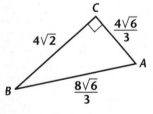

_____

8.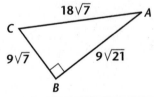

_____

9.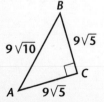

_____

10.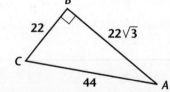

_____

Enrichment
8.2 The "Pythagorean" Theorem and the Unit Circle

The Pythagorean Theorem can be used with points on a unit circle. Notice that the length of the hypotenuse is 1 for the unit circle. Since $x = \cos \theta$ and $y = \sin \theta$, $x^2 + y^2 = c^2$ becomes $(\cos \theta)^2 + (\sin \theta)^2 = 1$.

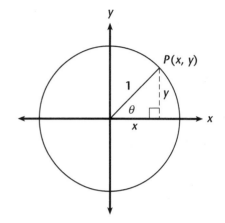

Use the Pythagorean Theorem to find the missing coordinate of a point on the unit circle in the first quadrant. Round answers to three decimal places.

1. $(0.7, y)$ _____

2. $(x, 0.42)$ _____

3. $(0.87, y)$ _____

4. $(x, 0.43)$ _____

5. $(x, 0.59)$ _____

6. $(0.21, y)$ _____

7. $(0.96, y)$ _____

8. $(0.07, y)$ _____

9. $(x, 0.88)$ _____

10. $(x, 0.65)$ _____

11. $(0.5, y)$ _____

12. $(x, 0)$ _____

Determine whether or not each point appears to be on the unit circle. (Allow for calculator rounding errors.)

13. $(0.6, 0.4)$ _____

14. $(0.32, 0.95)$ _____

15. $(-0.89, -0.46)$ _____

16. $(0.55, -0.38)$ _____

17. $(-0.47, 0.88)$ _____

18. $(1, -1)$ _____

19. $(0.94, -0.43)$ _____

20. $(0.63, 0.78)$ _____

21. $(-0.59, -0.81)$ _____

22. $(0.35, -0.94)$ _____

23. $(-1, 0.28)$ _____

24. $(-0.8, -0.7)$ _____

Enrichment

8.3 Trigonometric Functions of Inverse Trigonometric Functions

To find $\sin(\tan^{-1}1)$, first find $\tan^{-1}$, using the principal value. Since $\tan^{-1}1 = 45°$ and $\sin 45° = \frac{\sqrt{2}}{2}$ or ≈ 0.7071, $\sin(\tan^{-1}) = 0.7071$ to the nearest ten-thousandth.

Evaluate each expression. Round answers to the nearest ten-thousandth.

1. $\cos(\tan^{-1}0.8427)$

_____

2. $\sin(\cos^{-1}0.5)$

_____

3. $\tan(\sin^{-1}(-0.87))$

_____

4. $\cos(\cos^{-1}(-0.6374))$

_____

5. $\sin(\tan^{-1}8.7632)$

_____

6. $\cos(\sin^{-1}(-0.8762))$

_____

7. $\tan(\tan^{-1}(-4.6791))$

_____

8. $\cos(\sin^{-1}0.7426)$

_____

9. $\sin(\cos^{-1}0.8714)$

_____

10. $\cos(\tan^{-1}3.4217)$

_____

11. $\sin(\cos^{-1}0.0216)$

_____

12. $\cos(\cos^{-1}0.7111)$

_____

13. $\tan(\sin^{-1}0.2174)$

_____

14. $\tan(\cos^{-1}0.7111)$

_____

15. $\sin(\tan^{-1}0.4129)$

_____

16. $\cos(\sin^{-1}0.0027)$

_____

17. $\sin(\cos^{-1}(-0.5683))$

_____

18. $\tan(\sin^{-1}0.4579)$

_____

Enrichment
8.4 Two Equations for One Graph

For each graph, write one equation in terms of sin x and one
equation in terms of cos x.

1.

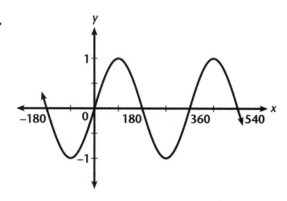

_____

2.

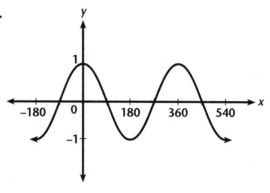

_____

3.

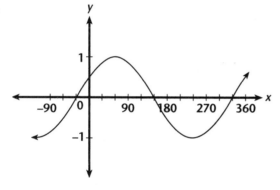

_____

4.

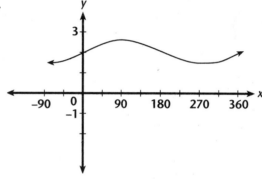

_____

5.

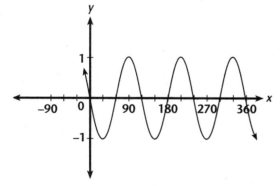

_____

6.

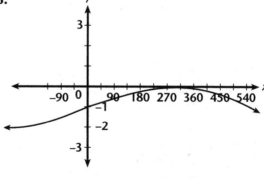

_____

Enrichment
8.5 Angular Speed

Radian measure can be used to describe the angular speed of
an object, or how fast the object is moving along a circular
path. Angular speed (ω) = is given by the formula $\omega = \frac{\theta}{t}$,
where θ = the angle through which the object moves, in
radians, and t = elapsed time.

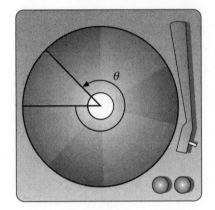

For example, before compact discs (CDs) were so popular you
could buy your favorite song on a "single" or "45" record. The
"45" meant that the record made 45 complete revolutions per
minute (rpm). Since 1 revolution = 2π radians,

$$45 \text{ rpm} = \frac{45 \text{ revolutions}}{\text{minute}} \times \frac{2\pi \text{ radians}}{\text{revolution}} = \frac{90\pi \text{ radians}}{\text{minute}}, \text{ or } 90\pi \text{ rad/min.}$$

Find the angular speed of each of the following objects.

1. a flywheel that turns 500 revolutions in 4 minutes _____

2. a record that turns $33\frac{1}{3}$ revolutions per minute _____

3. a mixer that turns at 1000 revolutions in 2 minutes _____

4. turning a flywheel at a rate of 60 revolutions in 45 seconds _____

5. a turntable that spins at 78 revolutions per minute _____

6. a potter's wheel that turns at 250 revolutions in 4 minutes _____

7. a wheel that turns 1080 times in 3 minutes _____

8. a coin that spins 25 times in 2 seconds _____

9. a beacon that turns 400 times in 80 minutes _____

10. a lariat that spins 45 times in 10 seconds _____

Enrichment
8.6 The Nautical Mile

A nautical mile is the distance on a great arc that is subtended by a central angle of 1 minute $\left(\frac{1}{60}\text{ of a degree}\right)$. To calculate the length of a nautical mile, a value of 3956 statute miles is used for the radius of the Earth. (1 statute mile = 5280 feet).

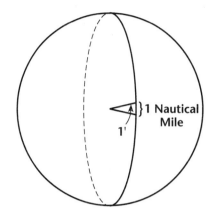

Find each measurement in statute miles. Round answers to the nearest hundredth.

1. 1 nautical mile   _____

2. 200 nautical miles _____

3. 48 nautical miles   _____

4. 576 nautical miles _____

5. 100 nautical miles _____

6. 50 nautical miles   _____

7. 82.3 nautical miles _____

8. 6.3 nautical miles   _____

9. 93.7 nautical miles _____

10. 4.8 nautical miles _____

Change each measurement to nautical miles.

11. 8 statute miles   _____

12. 2 statute miles   _____

13. 35 statute miles   _____

14. 75 statute miles   _____

15. 200 statute miles   _____

16. 2500 statute miles _____

17. 9500 feet   _____

18. 18,500 feet   _____

19. 100,000 feet   _____

20. 8,900,000 feet   _____

Enrichment

8.7 A Different Type of Motion

Projectile motion, such as the path of a thrown football, can be described by the function $f(t) = v_0(\sin \theta)t - 16t^2$, where t is the time in seconds, $f(t)$ is the height reached by the projectile, v_0 is the initial velocity of the projectile in ft/s, and θ is the angle at which the projectile is launched. If the metric system is used, $f(t)$ becomes $f(t) = v_0(\sin \theta)t - 4.9t^2$, where $f(t)$ is measured in meters and v_0 is the initial velocity in m/s.

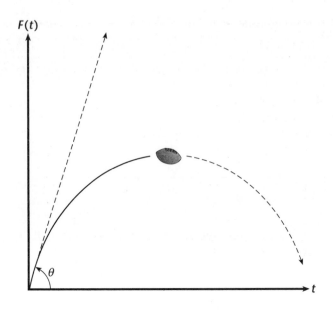

Find each answer to the nearest tenth.

1. A ball is thrown at an angle of 45° to the ground at an initial velocity of 80 ft/s. What is the height of the ball after 2 s?

_____

2. A rock is tossed at an angle of 60°, at an initial velocity of 50 m/s. What is the height of the rock after 1 s?

_____

3. A football is kicked at an initial velocity of 120 ft/s. When will the ball return to the ground if it was kicked at an angle of 30°?

_____

4. A soccer ball is kicked at an initial velocity of 40 m/s, at an angle of 45°. When will the ball be at a height of 10 m?

_____

5. A car hits a ramp at an initial velocity of 1.2 m/s. The ramp makes an angle of 15° with the ground. When will the car touch down on the other side?

_____

6. A skateboard hits a ramp at 20 ft/s, at an angle of 20°. What is the maximum height the skateboard will reach?

_____

Technology
8.1 Ratios in Special Right Triangles

The diagram shows a right triangle with two sides of equal length. In the diagram, $PQ = OQ$. This type of triangle is called an isosceles right triangle. From geometry, you know that the measures of the two acute angles are equal and equal to 45°.

You can use a spreadsheet like the one shown to study various relationships in this type of triangle. Use the formula SQRT(A2^2+B2^2) in cell C2.

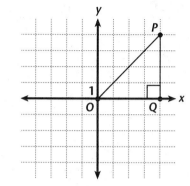

	A	B	C	D	E	F
1	OQ	PQ	OP	OQ/OP	PQ/OP	PQ/OQ
2	1	1				
3	2	2				
4	3	3				
5	4	4				
6	...	...				

Use a spreadsheet like the one shown.

1. From the spreadsheet, what can you say about the entries in columns

   D, E, and F? _____

One group of students decided that a right triangle is special if $PQ = nOQ$, where n is a positive integer. Modify the spreadsheet above accordingly for Exercises 2 through 7. What can you say about the entries in columns D, E, and F?

2. $n = 2$ _____

3. $n = 3$ _____

4. $n = 4$ _____

5. $n = 5$ _____

6. $n = 6$ _____

7. $n = 7$ _____

8. Suppose a right triangle is placed on a coordinate plane as shown above and PQ is some positive number times OQ. Do you think the pattern you saw develop in columns D, E, and F in Exercises 2–7 is still true? Explain.

   _____

   _____

Technology
8.2 Coterminal Angles and the Division Algorithm

In earlier studies, you learned that if you divide one number (the dividend) by another number (the divisor), you get a quotient and a remainder. For example, if you divide 124 by 12 you get a quotient of 10 and a remainder of 4. You can write this as shown.

$$124 = 10(12) + 4$$

In general, when you divide n by m, you get a quotient of q and a remainder of r, where $0 \le r < q$. This statement is known as the division algorithm. You can write this as shown.

$$n = qm + r, \text{ where } 0 \le r < q$$

You may be surprised to learn that you can use this idea and a spreadsheet to explore angles coterminal with an angle, such as 23°, in standard position.

	A	B	C
1	23		
2	113		
3	203		
4	293		
5	383		
6	...		

Cell A1 contains 23.
Cell A2 contains 90+A1.
Cell B1 contains INT(A1/360).
Cell C1 contains A1−360*B1.

1. Create the spreadsheet shown and fill the first 20 rows.

2. What does an entry in column B tell you about the corresponding number in column A?

   _____

3. List all angle measures in column A that have 23 as the corresponding entry in column C.

   _____

4. Write a formula that tells how your answers to Exercise 3 are related to 23.

   _____

5. Describe how to find the measure of r of the smallest angle of positive measure coterminal with an angle of positive measure s.

   _____

Technology
8.3 Solving Right Triangles Graphically

Suppose you want to find the values of a and b in Figure A. You can place the triangle on a coordinate system as drawn in Figure B. Finally, add a circle whose center is the origin with radius 20 as shown in Figure C.

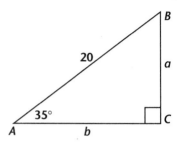

Figure A

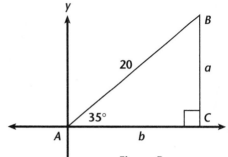

Figure B

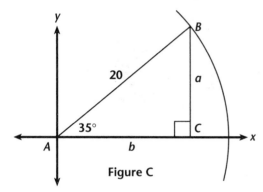

Figure C

The coordinates a and b of point B lie along the line through the origin with slope 35° and on the circle with radius 20 and center at the origin. Coordinates a and b satisfy $y = (\tan 35°)x$ and $x^2 + y^2 = 20^2$. The equation of the circle can be written $y = \sqrt{20^2 - x^2}$. Set the calculator in degree mode and use it to graph $y = (\tan 35°)x$ and $y = \sqrt{20^2 - x^2}$. over the interval $0 \le x \le 20$. Finally, use the trace feature to find the coordinates of intersection, $x \approx 16.38$ and $y \approx 11.47$. Thus, $a \approx 11.47$ and $b \approx 16.38$.

Use a graphics calculator to find a and b.

1. $m \angle A = 40°$; $AB = 15$

2. $m \angle A = 62°$; $AB = 15$

3. $m \angle A = 14°$; $AB = 23$

_____

_____

_____

4. $m \angle A = 80°$; $AB = 1.3$

5. $m \angle A = 10°$; $AB = 10$

6. $m \angle A = 73°$; $AB = 157$

_____

_____

_____

7. $m \angle A = 27°$; $AB = 5$

8. $m \angle A = 89°$; $AB = 8$

9. $m \angle A = 50°$; $AB = 243$

_____

_____

_____

Technology
8.4 The Graph of the Tangent Function

In Lesson 8.4, you explored the graphs of $y = \sin x$ and $y = \cos x$ and variations of those functions by using a graphics calculator. In this activity, you carry out a similar exploration for the function $y = \tan x$.

Use a graphics calculator in degree mode.

1. Graph $y = \tan x$ for $-360° \leq x \leq 360°$.

2. Briefly describe the calculator display in your own words.

_____

3. Write a formula for those values of x for which $\tan x = 0$.

_____

_____

4. Write a formula for those values of x for which $\tan x$ is undefined.

_____

_____

5. Find the domain and range of $y = \tan x$.

_____

_____

6. Is $y = \tan x$ periodic? What is its period?

_____

7. Does the tangent function have an amplitude? Explain.

_____

8. Find $\tan 1170°$, if it exists.

_____

HRW material copyrighted under notice appearing earlier in this work.

Technology
8.5 Fractions in Circles

Suppose that you sketch a central angle θ in a circle of radius 1 as shown in the diagram. From the diagram, you can see that the angle determines an arc on the circle. Conversely, if you sketch an arc on the circle, you determine a central angle θ. To make a connection between the angle measure and the length of the arc, reason as follows.

The angle measure θ is a certain fraction of 360°. It is reasonable to say that s is the same fraction of the circumference of the circle. Since $r = 1$, it may be omitted from the formula. Therefore:

$$\frac{\theta}{360°} = \frac{s}{2\pi} \text{ units}$$

You can now write two formulas that allow you to convert degrees to the corresponding radian measure and convert radian measure to degree measure.

$$s = \frac{\pi\theta}{180°} \approx 0.01745\theta \qquad \theta = \frac{180°s}{\pi} \approx 57.29578s$$

Use a scientific calculator to convert degree measure to radian measure or radian measure to degree measure.

1. 60°

2. 0.42 radians

3. 10°

_____   _____   _____

4. 1.6 radians

5. 72°

6. 2.00 radians

_____   _____   _____

Use your graphics calculator to graph $y = 0.01745x$, where y stands for s and x stands for θ over the interval $0 \le x \le 90$. Use the graph to convert degrees or radians to radians or degrees.

7. 30°

8. 0.8 radians

9. 45°

10. 1.1 radians

_____   _____   _____   _____

11. For what degree measures will the radian measure be more than 2π but less than 4π?

_____

Technology
8.6 Linear Speed

Suppose a point travels on a circle of radius 2.2 ft at 15.5 revolutions per minute. To find the distance traveled around the circle each minute, you need to evaluate $(2.2)(15.5)(2\pi) = 214.2566$. You can then say that the point is traveling around the circle at about 214.26 feet per minute.

The speed 15.5 revolutions per minute is called the angular speed and 214.26 feet per minute is called the linear speed of the point.

If you want to compute the linear speed of a point given a constant angular speed, you can use a spreadsheet. Enter a series of radii in column A, the given angular speed in column B, and the formula for the linear speed in column C. For example, cell C2 contains A2*B2*2*3.14159. (Note that 3.14159 is an approximation of π.)

If the radius is given in feet and the time unit for angular speed is given in minutes, you read the entry in column C as so many feet per minute.

	A	B	C
1	RADIUS	REV/MIN	LIN SPD
2	0.0		
3	0.5		
4	1.0		
5	1.5		
6	2.0		
7	...		

Given the information below, use the spreadsheet to find the linear speed.

1. radius = 0 to 5.0 in increments of 0.5 and angular speed 15.5 revolutions per minute

2. radius = 0 to 5.0 in increments of 0.5 and angular speed 24.2 revolutions per minute

3. radius = 22.7 ft and angular speed = 0 revolutions per minute to 6 revolutions per minute in increments of 0.5 revolutions per minute

4. radius = 130.2 ft and angular speed = 0 revolutions per minute to 6 revolutions per minute in increments of 0.5 revolutions per minute

5. If the angular speed is constant, what can you say about the linear speed as the radius of the circle increases?

_____

6. If the radius of the circle is constant and angular speed increases, what can you say about the linear speed of the point?

_____

Technology
8.7 Weights on Springs

Suppose that a weight is suspended from a spring and, when the experiment begins, the weight is pulled down 7 cm. If there are no forces that retard the motion of the weight, it will bob up and down in a periodic fashion. Suppose that the weight bobs in such a way that every 8 seconds it is at the lowest point in its motion.

You can make a table that shows the motion of the weight. If you use a graphics calculator to plot the points from the table, you will see that the motion appears to be sinusoidal.

Now experiment with different equations involving sine or cosine to see which equation goes through the points you plotted.

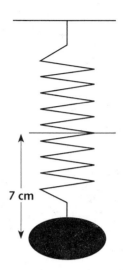

7 cm

time t	displacement y
0	7 cm down
2	0 cm up or down
4	7 cm up
6	0 cm up or down
8	7 cm down
10	0 cm up or down
12	7 cm up
14	0 cm up or down

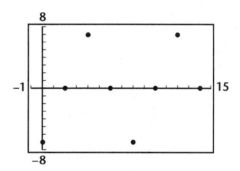

Find an equation that represents each motion.

1. At $t = 0$, the weight is pulled down 7 cm and it takes 8 seconds for the

   weight to return to its original position. _____

2. At $t = 0$, the weight is pushed up 7 cm and it takes 8 seconds for the

   weight to return to its original position. _____

3. At $t = 0$, the weight is pulled down 4.5 cm and it takes 6 seconds for the

   weight to return to its original position. _____

4. At $t = 0$, the weight is pushed up 4.5 cm and it takes 6 seconds for the

   weight to return to its original position. _____

5. At $t = 0$, the weight is pulled down a cm and it takes c seconds for the

   weight to return to its original position. _____

Lesson Activity
8.1 Modeling with Right Triangles

As shown, squares are drawn on each side of the 45-45-90 right triangle *ABC*. The vertices of the squares are connected to form triangles *IBD*, *FAE*, and *HCG*.

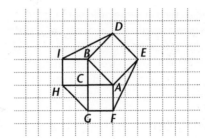

1. Find the area of each of the four triangles.

_____

2. How do the areas of the four triangles compare?

_____

Start with the 30-60-90 right triangle *ABC* as shown.

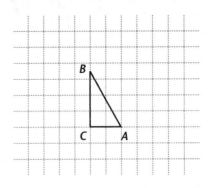

3. Draw a square on each side of triangle *ABC*. Label the vertices as in the diagram for Exercise 1. Then connect the vertices of the squares to form triangles *IBD*, *FAE*, and *HCG*.

4. Find the area of each of the four triangles.

_____

5. How do the areas of the four triangles compare?

_____

Start with the 3-4-5 right triangle *ABC* as shown.

6. Draw a square on each side of triangle *ABC*. Label the vertices of the squares, as in Exercise 3, and connect them to form triangles *IBD*, *FAE*, and *HCG*.

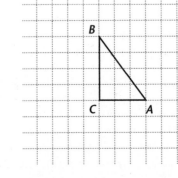

7. Find the area of each of the four triangles.

_____

8. How do the areas of the four triangles compare?

_____

9. What relationship do you see among the areas of the four triangles?

_____

Lesson Activity
8.2 Spinning Around a Unit Circle

Draw a unit circle on construction paper. Use the *x*- and
y-axes to mark the 90° angles. Mark where the terminal side
of each 30°, 45°, and 60° angle in standard position would
intersect the unit circle.

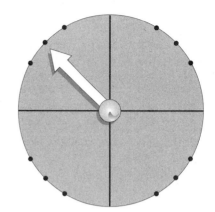

Make a spinner from construction paper, a cork, and a
thumb tack, and place it at the origin. Your spinner should
resemble the one shown.

1. From standard position, spin the spinner
    counterclockwise. Name a coterminal angle for the
    angle at which the spinner stops. _____

2. From standard position, spin the spinner clockwise.
    Name a coterminal angle for the angle at which the
    spinner stops. Name a second coterminal angle.

_____

**For Exercises 3–5, the common angles are placed into groups.
Each time you spin the spinner, record the marker the spinner is
nearest to when it stops.**

Marker	Tally	Frequency
0°, 90°, 180°, 270°		
30°, 150°, 210°, 330°		
45°, 135°, 225°, 315°		
60°, 120°, 240°, 300°		

3. Spin the spinner counterclockwise. When the spinner stops, move the
    spinner counterclockwise to the nearest marker. Record you answer in
    the tally section of the chart.

4. Repeat Exercise 3 and tally the results 24 more times.

5. To find the frequency, write a ratio comparing the number of times the
    spinner landed on each group of markers to the total number of spins.
    Find the sum of the ratios.

_____

6. Spin the spinner counterclockwise. When the spinner stops, move it

    counterclockwise to the nearest marked angle. Name the coordinates. _____

7. Repeat Exercise 6, spinning the spinner clockwise. Name the

    coordinates where the spinner stops. _____

_____

Lesson Activity
8.3 Approximating Trigonometric Functions with Polynomials

The measure of an angle in standard position can also be expressed in radians, where one radian is the measure of a central angle that intercepts an arc having the same length as the radius of the circle.

Computer programmers use certain polynomial functions to approximate the values of trigonometric functions. For example, the following polynomial can be used to approximate the sine of an angle measured in radians.

$$S(x) = x - \frac{x^3}{6} + \frac{x^5}{120} - \frac{x^7}{5040} + \frac{x^9}{362880} - \frac{x^{11}}{39916800}$$

Let d be the number of degrees in the given angle and let x be the radian measure of angle d. Use a calculator to compute the value of sin x using the polynomial. Then compare your results with the value of sin d obtained using your calculator in degree mode. The radian measures have been filled in for you.

	d (degrees)	x (radians)	$S(x)$	sin d
1.	6°	0.1047197551		
2.	46°	0.8028514559		
3.	154°	2.687807048		
4.	208°	3.630284884		

The polynomial $C(x) = 1 - \frac{x^2}{2} + \frac{x^4}{24} - \frac{x^6}{720} + \frac{x^8}{40320} - \frac{x^{10}}{3628800}$ can be used to approximate the cosine of an angle measured in radians. Complete the table.

	d (degrees)	x (radians)	$C(x)$	cos d
5.	6°	0.1047197551		
6.	46°	0.8028514559		
7.	154°	2.687807048		
8.	208°	3.630284884		

9. How can the tangent of any angle be approximated using polynomials?

_____

Lesson Activity
8.4 Periodic Functions on a Graphics Calculator

1. Using your calculator in degree mode, graph the function $f(\theta) = \sin \theta$. Use Xmin = 0, Xscl = 0, Ymin = -2, and Ymax = 2. Fill in the table for the number of full cycles on the graph of $f(\theta)$ as the value of Xmax change.

Xmax	360	34200	68040	101880	135720	169560
Cycles	1					

2. Using the same window as in Exercise 1, fill in the table for the number of full cycles on the graph of $f(\theta)$ as the values of Xmax change.

Xmax	33840	34200	34560	34920	35280	35640
Cycles						

The number of cycles pictured on a graphics calculator is related to the number of columns of pixels on the screen. For example, the TI-82 has 94 columns of pixels.

3. Find a formula to determine all the domains for which the graph of the sine function shows only one cycle.

   _____

4. Find a formula to determine all the domains for which the graph of the sine function has exactly n cycles.

   _____

Use your calculator in degree mode and dot mode to graph $f(\theta) = \cos \theta$. Set Xmin = 0, Xscl = 0, Ymin = -3, and Ymax = 3.

5. Change Xmax to 17280, then graph the function. Describe the graph. Change the mode to sequential, then graph the function. Compare the two graphs.

   _____

   _____

6. Change Xmax to 11640, then graph the function in dot mode. Describe the graph. Change the mode to sequential, then graph the function. Compare the two graphs.

   _____

   _____

Lesson Activity
8.5 Fixed Points and Cobwebs

Using your calculator in radian mode, graph the functions $f(\theta) = \cos \theta$ and $f(x) = x$. Use Xmin = 0, Xmax = 1.88, Xscl = 0, Ymin = −0.14, Ymax = 1.1. Record the y-coordinate for each point described below. Round your results to the nearest thousandths in the table.

Point	First	Second	Third	Fourth	Fifth	Sixth	Seventh
y							

1. For the first point, place the cursor on $x = 1.00$, $y = 0$. Choose Line from the DRAW menu and press Enter. Draw a vertical line to $f(\theta) = \cos \theta$ by moving the cursor vertically up. Press Enter when the cursor touches the graph of $f(\theta) = \cos \theta$.

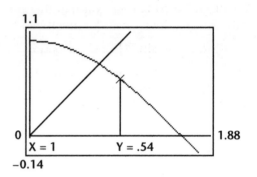

2. For the second point, press Enter twice and draw a horizontal line to $f(x) = x$. Press Enter twice again and draw a vertical line to the function $f(\theta) = \cos \theta$.

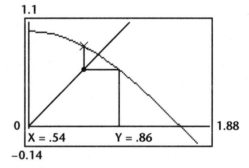

3. Press Enter twice and draw a horizontal line to $y = x$. Press Enter twice and draw a vertical line to $f(\theta) = \cos \theta$ for the third point.

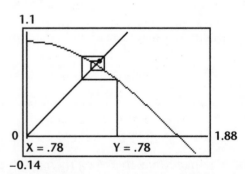

4. Repeat step 3 for the fourth, fifth, sixth, and seventh points.

5. Use your calculator to evaluate cos(1). Then evaluate cos(cos(1)). Next evaluate cos(cos(cos(1))). Continue evaluating cos(cos(cos(...(1)))). Compare your results to the values obtained in the table. Describe what you noticed.

_____

Lesson Activity
8.6 Starships, Arc Length, and Area

Suppose a starship is located at an altitude of 200 miles above the Earth. The straight line distance d, to the horizon is $d = \sqrt{2rh + h^2}$, where r is the radius of the Earth (approximately 3960 miles), and h is the altitude of the starship above the Earth. In this case, $d = \sqrt{2(3960)(200) + (200)^2} \approx$ 1274 miles.

The distance along the surface of the Earth is the length of arc CE and is given by the equation $CE = r\sin^{-1}\left[\dfrac{\sqrt{2rh + h^2}}{r + h}\right]$ radians where E is the point directly under the starship, C is the horizon, r is the radius of the Earth, and h is the altitude of the starship above the Earth.

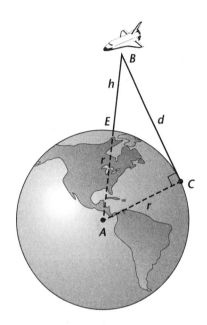

1. Find the straight line distance to the horizon when the starship is 100 miles above the Earth. Then find the distance for 150 miles.

    _____

2. Find the arc length along the surface of the Earth when the starship is 100 miles above the Earth. Then find the arc length for 150 and 200 miles.

    _____

3. What happens to the arc length as the distance of the starship above the Earth increases?

    _____

4. The portion of the Earth that can be seen from an altitude h, above the Earth is given by

    $$SA = \frac{2\pi r^2 h}{r + h},\ \text{where } r \text{ is the radius of the Earth.}$$

    Find the area of the Earth's surface that can be seen from an altitude of 200 miles above the Earth. Then find the area that can be seen from an altitude of 100 miles and 150 miles.

    _____

Lesson Activity
8.7 Application of Periodic Functions

A baseball is hit at a 20° angle from the ground in a straight line drive. The ball is 3 feet off the ground when it is hit and travels at a speed of 150 ft/s. Four hundred feet away is a 20-foot fence and the wind is blowing against the hitter at 9 mi/h. How can you determine if the ball is a home run?

Draw a picture, analyze the situation, and use parametric equations. Since distance = rate × time, $x = (150\cos 20°)t$ and $y = (150\sin 20°)t$. Since 60 mi/h = 88 ft/s, 9 mi/h = 13.2 ft/s.

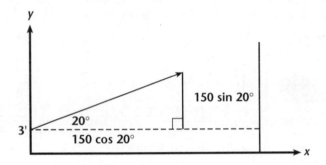

The total horizontal distance the ball travels is the horizontal distance − wind resistance. $X_{1T} = (150\cos 20°)T − 13.2$

The total vertical distance the ball travels is the vertical distance − gravity + initial height. $Y_{1T} = (150\sin 20°)T − 16T^2 + 3$

1. Set your calculator to parametric and degree mode. Use Tmin = 0, Tmax = 5, Tstep = 0.1, Xmin = 0, Xmax = 450, Xscl = 50, Ymin = 0, Ymax = 50, and Yscl = 10 as the range values. Then graph the equations.

2. Use trace to find the height at $x = 400$. Does the ball clear the fence?

_____

3. Where is the ball 2.8 seconds after it is hit? Can an outfielder catch the ball?

_____

4. Graph the fence on the screen using parametric equations. Since the fence is 400 feet away from the pitcher, $X_{2T} = 400$. Since the fence is 20 feet high, $Y_{2T} = 20(T/Tmax)$. Note that Tmax can be found in the VARS menu.

5. Does the ball traveling at 150 ft/s hit the fence, if there is no wind resistance?

_____

6. Does the ball traveling at 160 ft/s hit the fence if the wind resistance is 9 mi/h?

_____

Assessing Prior Knowledge
8.1 Exploring Special Right Triangles

Let *a* and *b* be the legs of a right triangle and *c* its hypotenuse. Find

1. c if $a = 1, b = 1.$ _____

2. b if $a = 1, c = 2.$ _____

- -

NAME _____ CLASS _____ DATE _____

Quiz
8.1 Exploring Special Right Triangles

Find *x* and *y*.

1.
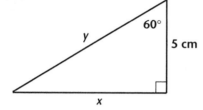

2.
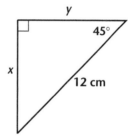

_____   _____

3.
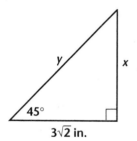

4.
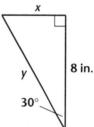

_____   _____

5. Find x.
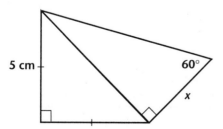

6. Find the area of the figure.
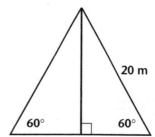

_____   _____

Assessing Prior Knowledge
8.2 The Unit Circle

Find the distance between the following pairs of points.

1. (2, 3) and (4, 8) _____

2. (0, 0) and $\left(\frac{\sqrt{2}}{2}, \frac{\sqrt{2}}{2}\right)$ _____

3. (0, 0) and $\left(\frac{1}{2}, \frac{\sqrt{3}}{2}\right)$ _____

- -

Quiz
8.2 The Unit Circle

1. Express one-third of a full rotation in degrees. _____

Draw each angle in standard position.

2. 148°

3. 50°

4. −320°

5. −75°

6. −800°

7. 650°

Use your calculator to find *x* and *y* to the nearest hundredth.

8.

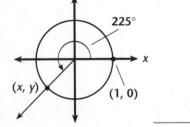

_____

9.

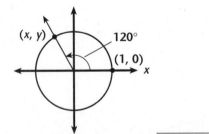

_____

10.

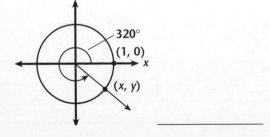

_____

11.

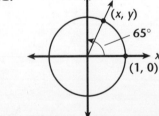

_____

Assessing Prior Knowledge
8.3 Trigonometric Applications

Find the inverse of each function.

1. $f(x) = 3x - 8$ _____

2. $f(x) = 2e^x$ _____

- -

NAME _____ CLASS _____ DATE _____

Quiz
8.3 Trigonometric Applications

Find two angles between 0° and 360° for each of the following inverse relations.

1. $\tan^{-1}\left(\dfrac{\sqrt{3}}{3}\right)$ _____

2. $\cos^{-1}\left(\dfrac{1}{2}\right)$ _____

3. $\sin^{-1}\left(\dfrac{\sqrt{3}}{2}\right)$ _____

4. $\tan^{-1}\left(-\sqrt{3}\right)$ _____

5. How do you find the measure of the angle whose terminal side intersects the unit circle at the point (0.92, 0.40)?

_____

Can each statement possibly be true? Explain your answers.

6. $\cos^{-1}\left(\dfrac{\sqrt{3}}{2}\right) = \sin^{-1}\left(-\dfrac{1}{2}\right)$ _____

7. $\sin^{-1} 0 = \cos^{-1} 0$ _____

8. Find all values of $\tan\theta$ for which $\sin\theta = \cos\theta$.

_____

NAME _____ CLASS _____ DATE _____

Assessing Prior Knowledge
8.4 Exploring the Sine and Cosine Graphs

Find the vertex of the parabola described by each quadratic function.

1. $f(x) = (x + 1)^2 - 2$ _____

2. $f(x) = -(x - 3)^2 + 4$ _____

- -

NAME _____ CLASS _____ DATE _____

Quiz
8.4 Exploring the Sine and Cosine Graphs

**Identify the vertical shift, amplitude, period, and phase shift
for each function. Then graph the function.**

1. $f(\theta) = \sin 4\theta$

_____

2. $g(\theta) = 3 + \sin \theta$

_____

3. $h(\theta) = -5\sin 2\theta$

_____

4. $j(\theta) = -2 + 3 \cos \frac{1}{2}(\theta + 90°)$

_____

Write a function of the form $f(\theta) = a + b\sin c(\theta - d)$ for each graph.

5.

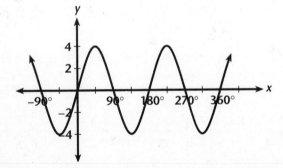

_____

6.

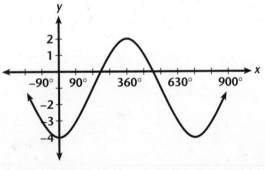

_____

Mid-Chapter Assessment
Chapter 8 (Lessons 8.1 – 8.4)

Write the letter that best answers the question or completes the statement.

_____ **1.** What is the value of x in the given diagram?

       a. 14 cm   **b.** 3.5 cm

       c. $7\sqrt{3}$ cm   **d.** $7\sqrt{2}$ cm

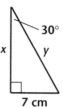

_____ **2.** The value of y to the nearest hundredth is

       a. −0.50   **b.** −0.87

       c. 0.87   **d.** 0.50

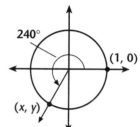

_____ **3.** One angle for which $\cos^{-1}\left(-\dfrac{\sqrt{3}}{2}\right)$ is true is

       a. 330°   **b.** 30°   **c.** 150°   **d.** 120°

_____ **4.** If $f(\theta) = -1 + 2\cos 3(\theta - 90°)$, what is the period of $f(\theta)$?

       a. 2   **b.** 90°   **c.** 3   **d.** 120°

5. Find y.

_____

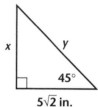

6. Find the area of the given figure.

_____

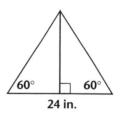

7. Draw −150° in standard position.

8. Sketch the graph of $g(\theta) = -2\sin 3\theta$.

Assessing Prior Knowledge

8.5 Radian Measure

Use a calculator to compute each value.

1. $\dfrac{360}{2\pi}$ _____

2. $\dfrac{2\pi}{360}$ _____

- -

Quiz

8.5 Radian Measure

Give the exact value for each of the following.

1. $\sin \dfrac{\pi}{3}$ _____

2. $\tan \dfrac{3\pi}{2}$ _____

3. $\cos \dfrac{5\pi}{4}$ _____

4. $\sin 3\pi$ _____

Convert degree measure to radian measure.

5. $70°$ _____

6. $-160°$ _____

Draw the angle with the given measure of rotation, and convert radian to degree measure.

7. $-\dfrac{3\pi}{2}$ _____

8. 1.5 _____

Identify the vertical shift, amplitude, period (in radians), and phase shift for each function.

9. $f(x) = 3 \sin \dfrac{\pi}{2} x$ _____

10. $f(x) = -2 - \cos \dfrac{1}{3}\left(x + \dfrac{\pi}{3}\right)$ _____

Using your calculator in radian mode, find each value.

11. $\sin 0.5$ _____

12. $\cos \pi$ _____

Assessing Prior Knowledge
8.6 Arc Length and Sector Area

Find the circumference of a circle with the following radii.

1. 1 _____   **2.** 1.5 _____   **3.** 2.3 _____

- -

Quiz
8.6 Arc Length and Sector Area

1. Find the length of $\widehat{AB}$.

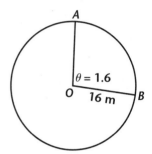

_____

2. Find the area of sector *COD*.

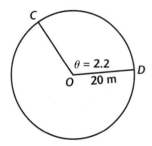

_____

3. Find the area of the shaded sector.

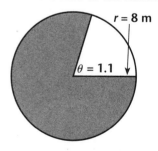

_____

4. Find the length of $\widehat{GH}$.

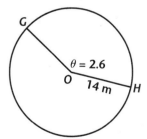

_____

A model plane attached to a 20-foot wire is controlled by a hobbyist who holds the handle at the opposite end of the wire and pivots about a fixed point.

5. Find the distance the plane travels if it covers an arc intercepted by a central angle of 115° in 4.2 seconds.

_____

6. Find the velocity of the plane in feet per second.

_____

Assessing Prior Knowledge
8.7 Applications of Periodic Functions

In the following function $f(x) = 5 + 2 \cos 3(x + \pi)$, identify the vertical shift, horizontal phase shift, amplitude, and period.

_____

- -

Quiz
8.7 Applications of Periodic Functions

Use a calculator in radian mode to graph the function
$f(x) = 3 + 5 \cos\left(\frac{x\pi}{2}\right).$

1. Find the maximum and minimum points on the graph. _____

2. Find $f(3)$. _____

3. Find $f(9)$. _____

4. Find two values of x such that $f(x) = 8$. _____

5. Find the period of the function. _____

6. Find the amplitude of the function. _____

7. Determine the vertical shift of the function. _____

Chapter Assessment
Chapter 8, Form A, page 1

Write the letter that best answers the question or completes the statement.

_____ **1.** The value of x in $\triangle ABC$ is

    a. $\dfrac{5\sqrt{3}}{2}$ cm   **b.** 2.5 cm

    c. $\dfrac{5\sqrt{2}}{2}$ cm   **d.** $5\sqrt{3}$ cm

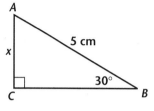

_____ **2.** The value of y in $\triangle RST$ is

    a. $\dfrac{7\sqrt{3}}{2}$ m   **b.** 3.5 m

    c. $\dfrac{7\sqrt{2}}{2}$ m   **d.** $7\sqrt{2}$ m

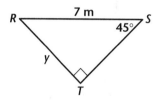

_____ **3.** What is the value of x to the nearest hundredth?

    a. 0.64   **b.** −0.77
    c. −0.64   **d.** 0.77

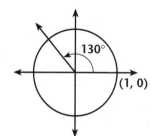

_____ **4.** Which angle is coterminal with the angle shown?

    a. 65°   **b.** 55°
    c. −235°   **d.** −145°

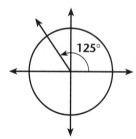

_____ **5.** An angle having the inverse relation $\cos^{-1}\left(-\dfrac{1}{2}\right)$ is

    a. −30°   **b.** 120°   **c.** 210°   **d.** −60°

_____ **6.** If the coordinates of the point where θ intersects the unit circle are (0.68, 0.73), what is the measure of θ to the nearest degree?

    a. 47°   **b.** 313°   **c.** 227°   **d.** 133°

_____ **7.** What is the amplitude of the function $f(x) = -3 + 4\cos\frac{1}{2}(x + 180°)$?

      a. $\frac{1}{2}$ **b.** 4   **c.** -3 **d.** 1

_____ **8.** What function is represented by the graph shown?

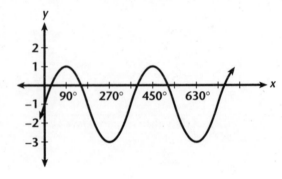

      a. $-1 + 2\cos\theta$ **b.** $1 + 2\sin2\theta$ **c.** $-1 - 2\cos\theta$ **d.** $-1 + 2\sin\theta$

_____ **9.** The value of $\frac{7\pi}{6}$ is

      a. $-\frac{1}{2}$ **b.** $-\frac{\sqrt{3}}{2}$ **c.** $\frac{1}{2}$ **d.** $\frac{\sqrt{3}}{2}$

_____ **10.** In $\triangle RST$, the value of x in radians is

      a. 0.72   **b.** 0.88
      c. 0.51   **d.** 1.07

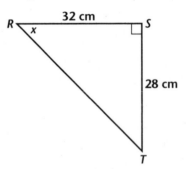

_____ **11.** What is the area of sector AOB?

      a. 31.2 m^2 **b.** 15.6 m^2
      c. 40.56 m^2 **d.** 187.2 m^2

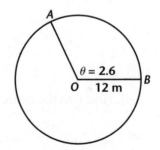

_____ **12.** What is the length of $\overarc{RS}$?

      a. 27.5 m   **b.** 6.05 m
      c. 5.5 m   **d.** 2.75 m

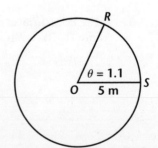

Chapter Assessment
Chapter 8, Form B, page 1

1. Find the value of x.

_____

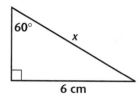

2. Find the area of the given figure.

_____

3. Draw $-510°$ in standard position.

4. Use a calculator to find x to the nearest hundredth.

_____

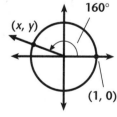

5. Let the coordinates of the point where θ intersects the unit circle be

$(-0.8, -0.6)$. Find θ to the nearest degree. _____

6. Find two angles between 0° and 360° for the inverse relation

$\cos^{-1}\left(-\dfrac{\sqrt{3}}{2}\right)$. _____

7. Find one angle for which $\cos^{-1}\left(-\dfrac{1}{2}\right) = \sin^{-1}\left(-\dfrac{\sqrt{3}}{2}\right)$ is true. _____

8. Identify the vertical shift, amplitude, period, and phase shift for the function $f(x) = -1 + 2\sin3\left(x - \frac{\pi}{2}\right)$.

_____

9. Write the function of the form $f(\theta) = a + b\sin c\,(\theta - d)$ for the given graph.

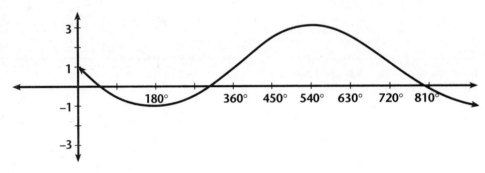

_____

10. Give the exact value of $\cos 5\pi$. _____

11. Draw the angle with rotation measure $-\frac{5\pi}{4}$.

12. Use a calculator in radian mode to find $\cos\frac{2\pi}{5}$ to the nearest hundredth. _____

13. Find the length of $\overset{\frown}{CD}$.

_____

14. Use a calculator in radian mode to graph the function $f(x) = 2 + 3\cos\left(\frac{x\pi}{4}\right)$. Find the first two values of x such that $f(x)$ is 0.

_____

Alternative Assessment
Exploring Graphs of Trigonometric Functions, Chapter 8, Form A

TASK: To graph the sine and the cosine functions

HOW YOU WILL BE SCORED: As you work through the task, your teacher will be looking for the following:

- whether you can identify the period, amplitude, phase shift and vertical shift of the sine function and the cosine function
- how well you can write the sine function and the cosine function

Graph each function with domain in degree measure, and find the period, amplitude, and phase shift.

1. $f(x) = \cos(x - 45°)$ _____

2. $g(x) = 2\sin(x + 60°)$ _____

3. Predict the period, amplitude, and vertical shift for $h(x) = 2 + \sin(x - 30°)$.

_____

4. Let $f(x) = (\cos - d)$ and $f(x) = \sin(x - d)$. Describe the effects of d on these functions.

_____

Graph each function with domain in degree measure, and find the domain, range, and period.

5. $f(x) = \cos 2(\theta - 45°)$ _____

6. $g(x) = \sin \frac{1}{2}(\theta + 60°)$ _____

7. Predict the domain, range, and period of $h(x) = 5\sin \frac{1}{3}(\theta - 30°)$.

_____

8. Write a function in the form $f(x) = a + b \sin c(x + d)$ with a degree measure domain of all real numbers, range $-1 \leq y \leq 1$, and period $120°$.

_____

SELF-ASSESSMENT: Compare the graphs of the function $f(x) = a - b \cos c(x - d)$ to the graph of the function $g(x) = a + b \cos c (x - d)$. How are they alike? How are they different?

Alternative Assessment
Applications of Trigonometric Functions, Chapter 8, Form B

TASK: To use graphs of trigonometric functions to model real-world situations

HOW YOU WILL BE SCORED: As you work through the task, your teacher will be looking for the following:

- whether you can use trigonometric functions to solve real-world problems
- how well you can communicate your responses in writing

Blood pressure (BP) is a measurement of the pressure that the blood exerts on the walls of the arteries during various stages of activity. Blood pressure is measured in millimeters of mercury on an instrument known as a sphygmomanometer. Diastolic pressure is the constant pressure that is in the walls of the arteries when the heart is at rest or between contractions.

The function $f(x) = 100 - 20 \cos 5\pi\frac{x}{3}$ approximates the BP, $f(x)$, in millimeters of mercury at time x in seconds for a person at rest.

Use a graphics calculator in degree mode to answer the following questions.

1. Find the period of the function. What does it represent?

   _____

2. Find the diastolic pressure when the mercury reading is 1.5 mm.

   _____

3. Find two times when the diastolic pressure is 110.

   _____

4. Identify the amplitude of the function and explain what is represents.

   _____

5. Identify the vertical shift of the function.

   _____

SELF-ASSESSMENT: What are some factors present in individual persons that can influence the blood pressure readings?

Practice & Apply
9.1 Inverse Variation

In the table, y varies inversely as the cube of x.

x	2	4	10
y	0.5	____	____

1. Find the constant of variation. _____

2. Write the equation of variation. _____

3. Complete the table. Then check your answers using the statistical regression and trace features of your graphics calculator.

4. If y varies inversely as x, complete the table.

x	3	6	____	a	$12a$
y	4	____	1.5	$\dfrac{12}{a}$	$\dfrac{1}{a}$

Find the constant of variation and write the equation of variation.

5. y varies inversely as x, and y is 8 when x is 1. _____

6. y varies inversely as the square of x, and y is 8 when $x = 1$. _____

7. y varies inversely as the cube of x, and y is 8 when $x = 2$. _____

A surveyor's 100-foot-long steel tape measure weighs approximately 0.0163 pounds per foot. The correction for the sag in the tape measure placed on a surface varies inversely with the square of the force (in pounds) with which it is pulled. If a distance of 75 feet is to be marked off with the tape measure, the constant of variation is 4.67.

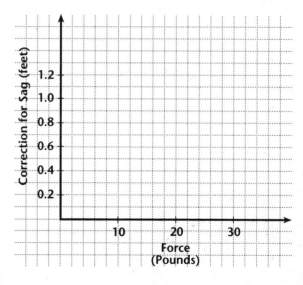

8. Find the variation equation. Then sketch its graph on the grid provided.

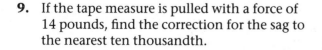

9. If the tape measure is pulled with a force of 14 pounds, find the correction for the sag to the nearest ten thousandth.

   _____

10. As the force increases, what happens to the correction for the sag?

   _____

Practice & Apply
9.2 Simple Reciprocal Functions

Match each graph with its function.

_____ **1.** 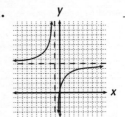   _____ **2.** 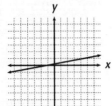   _____ **3.** 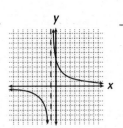   _____ **4.**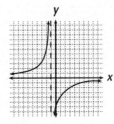

A. $y = \dfrac{6}{x+1}$ **B.** $y = \dfrac{-6}{x+1}$ **C.** $y = \dfrac{x+1}{6}$ **D.** $y = \dfrac{6x}{x+1}$

Use your graphics calculator to graph each function. Identify the domain and range for each function and find its asymptotes.

5. $f(x) = \dfrac{1}{x-1}$ _____

6. $f(x) = \dfrac{x-1}{x}$ _____

7. $f(x) = 1 + \dfrac{1}{x}$ _____

8. $f(x) = \dfrac{x-1}{x+1}$ _____

A gardener mixes pure water with 5 liters of a 10% weed killer solution.

9. Write the concentration of weed killer C, as a function of the amount of pure water added x.

_____

10. Graph the function on the grid provided.

11. What is the concentration when 5 liters of water has been added?

_____

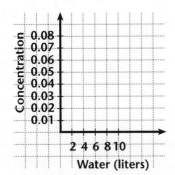

Practice & Apply
9.3 Exploring Reciprocals of Polynomial Functions

Determine whether each function is even, odd, or neither.

1. $f(x) = \dfrac{1}{x^2 - 1}$ _____

2. $f(x) = \dfrac{1}{x^3 - x}$ _____

3. $f(x) = \dfrac{1}{x^3 - 1}$ _____

Use the grid provided to sketch the graph of the reciprocal of each polynomial function.

4. $f(x) = x^2 + 8x + 15$

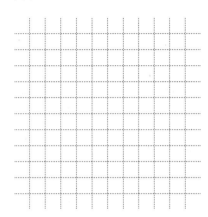

5. $f(x) = x^2 - 4x - 32$

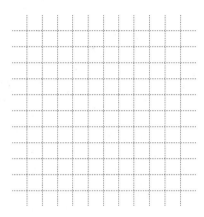

Assume that average annual copper prices in the world market are modeled by the equation $f(x) = \dfrac{1}{0.91 + 0.003x^2}$ where $x = 0$ corresponds to 1991 and $f(x)$ is the price of copper per pound.

6. Graph the function on the grid provided.

7. What is the domain of the function?

_____

8. Compare the average annual prices of copper per pound in 1991 and 1994.

_____

9. What is the projected average annual price of copper in 1999?

_____

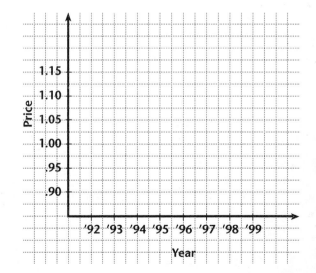

Practice & Apply
9.4 Quotients of Polynomial Functions

Simplify each rational function.

1. $\dfrac{x^2 - 16}{x^2 - 7x + 12}$ _____

2. $\dfrac{2x^2 + 11x + 5}{2x^2 + 13x + 15}$ _____

3. $\dfrac{-2x}{x^2 - x}$ _____

4. $\dfrac{x^2 + 7x + 10}{x^3 + 2x^2}$ _____

Find the vertical and horizontal asymptotes for each function and give the domain.

5. $\dfrac{x^2 - 4x - 12}{x^2 + 10x + 24}$ _____

6. $\dfrac{x^2 + 6x + 5}{(x - 2)(x + 1)}$ _____

7. $\dfrac{x^2 + 3x}{x^3 + 3x - 4x}$ _____

8. $\dfrac{x^2 - 4x + 4}{x^2 - 2x}$ _____

Assume that the average purchasing power of the dollar, in terms of the 1967 dollar, is modeled by the equation $f(x) = \dfrac{1000}{(x + 10)(x + 109)}$ where x represents the number of years since 1970. If $x = 0$ corresponds to 1970, find each of the following.

9. $f(0)$ _____   **10.** $f(5)$ _____   **11.** $f(10)$ _____   **12.** $f(20)$ _____   **13.** $f(30)$ _____

14. What does the horizontal asymptote mean about the average purchasing power of the dollar?

_____

Practice & Apply
9.5 Solving Rational Equations

Simplify each rational expression.

1. $\dfrac{2}{x+5} + \dfrac{x}{x-5}$ _____

2. $\dfrac{1}{x-5} + \dfrac{1}{x^2-25}$ _____

3. $\dfrac{1}{x} - \dfrac{x}{x+5}$ _____

4. $\dfrac{x+5}{x} - \dfrac{1}{x+5}$ _____

Use a graphics calculator to approximate all solutions of each equation.

5. $\dfrac{3x+8}{x^2+2x} = 2$ _____

6. $\dfrac{9x-4}{x^2+4x} = 1$ _____

7. $\dfrac{5x^2-6x}{x+1} = 4$ _____

8. $\dfrac{x-4}{2x+1} = \dfrac{1}{3}$ _____

9. Two numbers are said to be in the Golden Ratio when the ratio of the smaller to the larger number is approximately 0.618. If a line segment 12 meters long is to be divided using the Golden Ratio, find the length of each segment to the nearest hundredth.

_____

Solve each rational equation by algebraic means.

10. $\dfrac{3}{x-2} - \dfrac{2x}{x^2-4} = \dfrac{5}{x+2}$

11. $\dfrac{3x}{x+2} + \dfrac{4x+6}{x} = \dfrac{33}{x^2+2x}$

_____   _____

12. The cost, in millions of dollars, for the Drug Enforcement Administration to seize a percentage of domestic heroin, cocaine, marijuana, and dangerous drugs is modeled by the equation $f(x) = \dfrac{528x}{100-x}$ where x is the percentage of drugs removed. Find the percentage of drugs removed, if the cost is $166,000,000.

_____

Enrichment
9.1 Other Variation Functions

You have studied direct variation and inverse variation functions. Each of these functions can vary with the square or cube of a quantity and many also vary with more than one variable. Suppose, for example, y varies directly as the square of x and inversely as z. If $y = 36$ when $x = 2$ and $z = 4$, how can you find y when $x = 4$ and $z = -6$? First, set up a general equation that expresses the relationships among x, y, and z.

y varies directly as the square of x and inversely as z.

$$y = k\frac{x^2}{z}$$

Substitute the known values to find the constant of variation.

$$36 = k\frac{2^2}{4}$$

Solve for k.

$$36 = k$$

Rewrite the general equation using this value of k.

$$y = 36\frac{x^2}{z}$$

Substitute the second set of known values.

$$y = 36\frac{4^2}{-6}$$

Solve for y.

$$y = -96$$

Solve each variation problem.

1. y varies directly as x and z. $y = 24$ when $x = 2$ and $z = 3$. Find y when $x = 2$ and $z = 6$.   _____

2. y varies inversely with the square of x. $y = 24$ when $x = 4$. Find x when $y = 12$.   _____

3. y varies directly as the cube of x. $y = 8$ when $x = 6$. Find y when $x = 12$.   _____

4. y varies inversely as the cube of x. $y = 6$ when $x = 2$. Find y when $x = 8$.
                                        _____

5. y varies directly as the square root of x. $y = 8$ when $x = 36$. Find y when $x = 64$.   _____

6. y varies directly as z and inversely as x and w. $y = 36$ when $z = 2$, $w = 4$, and $x = 3$. Find y when $z = 3$, $w = 6$, and $x = 9$.   _____

7. y varies directly as x and w and inversely as the square of z. $y = 24$ when $w = 8$, $x = 3$, and $z = 2$. Find y when $w = 3$, $x = 4$, and $z = 6$.   _____

8. y varies inversely as the square of x and the cube of z. $y = 64$ when $x = 2$ and $z = 4$. Find y when $x = 4$ and $z = 2$.   _____

Enrichment
9.2 Holes in Graphs

Some functions can be rewritten as reciprocal functions. The graphs of the two functions will be identical except that the original function will have a hole at an undefined point.

For example, to graph $y = \dfrac{x + 2}{x^2 + 5x + 6}$, factor the denominator and simplify.

$y = \dfrac{x + 2}{x^2 + 5x + 6} = \dfrac{x + 2}{(x + 2)(x + 3)} = \dfrac{1}{x + 3}$. As you can see, by comparing these graphs,

the graph of $y = \dfrac{x + 2}{x^2 + 5x + 6}$ is identical to the graph of $y = \dfrac{1}{x + 3}$ with one exception;

is undefined at $x = -2$, there is a hole in the graph at that point.

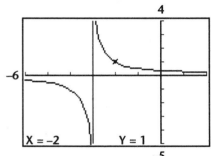  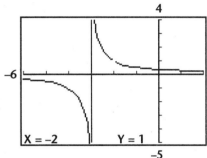

For each of the following functions, (a) find the equation of the vertical asymptote, (b) find the x-coordinate of the undefined point, and (c) use reciprocal functions to sketch the graph on the grid provided.

1. $y = \dfrac{x + 3}{x^2 - x - 12}$

   _____

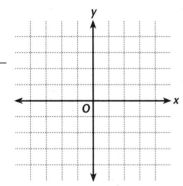

2. $y = \dfrac{x - 4}{x^2 + x - 20}$

   _____

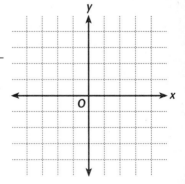

3. $y = \dfrac{x - 6}{x^2 - 7x + 6}$

   _____

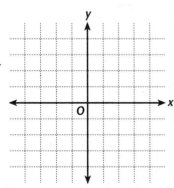

4. $y = \dfrac{x + 2}{x^2 + 3x + 2}$

   _____

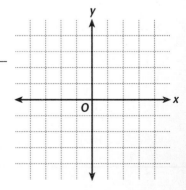

Enrichment
9.3 Problems Using Reciprocal Functions

Greg and Marti are planning a 650-mile trip. Find the time the trip will take at each average speed, if time $= \dfrac{650}{\text{average speed}}$.

1. 50 mi/h _____

2. 55 mi/h _____

3. 60 mi/h _____

4. 40 mi/h _____

5. 65 mi/h _____

6. 45 mi/h _____

Acceleration is defined as force divided by mass. In the metric system, force f, is measured in Newtons. Mass m, is measured in kilograms; and acceleration a, is measured in meters per second squared, m/s². For a force of 1 Newton, find the acceleration of each mass.

7. 100 kg _____

8. 1 kg _____

9. 50 kg _____

10. 4 kg _____

11. 0.01 kg _____

12. 0.001 kg _____

When a spring is stretched, the spring constant is defined as $k = \dfrac{4\pi^2 m}{T^2}$.

Here, m is the mass of the spring in kilograms and T is the period, or time it takes for the spring to move from its original position and return to that original position. Find k for a spring that has a mass of 1 kg and the period T (in seconds) is as given.

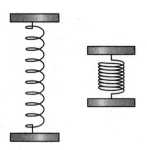

13. 1 s _____

14. 0.5 s _____

15. 2 s _____

16. 0.25 s _____

17. 0.01 s _____

18. 10 s _____

Electrical appliances offer a resistance or opposition to the flow of electricity within an electric circuit. For several appliances hooked up in parallel, the total resistance, called the equivalent resistance, R_{eq}, can be found by using the equation $\dfrac{1}{R_{eq}} = \dfrac{1}{R1} + \dfrac{1}{R2} + \dfrac{1}{R3}$, where R1, R2, and R3 are the resistances of the individual appliances. Find the total resistance for each set of resistances hooked up in parallel. Round answers to the nearest hundredth.

19. 40, 40 40 _____

20. 75, 100, 20 _____

21. 50, 75, 100 _____

22. 20, 30, 120 _____

23. 100, 80, 40 _____

24. 20, 60, 100 _____

Enrichment

9.4 The Oblique Asymptote

One more type of asymptote exists that is neither a vertical nor a horizontal line. This asymptote is called an oblique asymptote. An oblique symptote occurs when the greatest exponent of the polynomial in the numerator is larger than the greatest exponent of the polynomial in the denominator.

Consider the function $y = \dfrac{x^2 + 7x + 12}{x + 1}$. The denominator gives a vertical asymptote at $x = -1$. Since the greatest exponent of the numerator (the 2 of x^2) is greater than the greatest exponent of the denominator (the 1 of x), the function has an oblique asymptote. Divide to find its equation.

$$x + 1 \overline{)x^2 + 7x + 12} \quad \begin{array}{c} x + 6 + \dfrac{6}{x + 1} \end{array}$$

As x becomes larger, $\dfrac{6}{x + 1}$ approaches 0, so the equation of the oblique asymptote is $y = x + 6$. The graph of the function is shown.

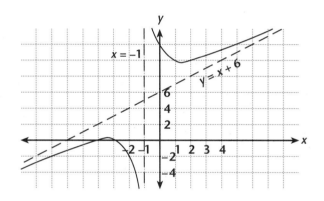

For each function, determine whether the nonvertical asymptote is horizontal or oblique and find its equation.

1. $y = \dfrac{2x + 3}{x + 4}$ _____

2. $y = \dfrac{x^2 + 3x + 1}{x - 2}$ _____

3. $y = \dfrac{2x + 7}{x^2 + 4x - 1}$ _____

4. $y = \dfrac{3x + 1}{2x - 8}$ _____

5. $y = \dfrac{3x^2 - 2x + 6}{x + 4}$ _____

6. $y = \dfrac{2x + 7}{x^2 + 4x - 1}$ _____

7. $y = \dfrac{x^2 - 3x + 4}{x^2 + x - 6}$ _____

8. $y = \dfrac{3x - 8}{x + 4}$ _____

9. $y = \dfrac{4}{2x - 8}$ _____

10. $y = \dfrac{2x - 7}{3x^2 + 2x - 6}$ _____

11. $y = \dfrac{2x^2 - 3x + 6}{2x + 1}$ _____

12. $y = \dfrac{5x^2 - 2x + 1}{x - 4}$ _____

Enrichment
9.5 Solving Rational Inequalities

To solve a rational inequality such as $\dfrac{x}{x-2} + \dfrac{2x}{x+2} > \dfrac{5}{x^2-4}$, first multiply by the least common denominator, to obtain an equivalent inequality. Use this inequality to divide a number line into intervals, then test points within each interval. Be sure to include any values for which the original inequality is undefined in determining your intervals.

$$(x-2)(x+2)\left(\dfrac{x}{x-2} + \dfrac{2x}{x+2}\right) > (x-2)(x+2)\left(\dfrac{5}{x^2-4}\right)$$

$$3x^2 - 2x - 5 > 0$$
$$(3x-5)(x+1) > 0$$

Here the zeros are $x = \dfrac{5}{3}$ and $x = -1$, and the undefined points of the original inequality are $x = 2$ and $x = -2$. Test the points in the original inequality.

$x < -2 \qquad -2 < x < -1 \qquad\qquad -1 < x < \dfrac{5}{3} \qquad \dfrac{5}{3} < x < 2 \qquad x > 2$

Interval	Test Point	Test	Result
$x < -2$	-3	$\dfrac{33}{5} > 1$	True
$-2 < x < -1$	$-\dfrac{3}{2}$	$-\dfrac{39}{7} > -\dfrac{20}{7}$	False
$-1 < x < \dfrac{5}{3}$	0	$0 > -\dfrac{5}{4}$	True
$\dfrac{5}{3} < x < 2$	$\dfrac{11}{6}$	$-\dfrac{231}{23} > -\dfrac{180}{23}$	False
$x > 2$	3	$\dfrac{21}{5} > 1$	True

So, the solution is $x < -2$ or $-1 < x < \dfrac{5}{3}$ or $x > 2$.

Find the solution of each inequality.

1. $\dfrac{x+1}{x-3} > 4$ _____

2. $\dfrac{x-2}{x+6} > 3$ _____

3. $\dfrac{x-2}{x+2} < 3$ _____

4. $\dfrac{x-5}{x+1} < 2$ _____

5. $\dfrac{x}{x+1} - \dfrac{2}{x-1} > 1$ _____

6. $\dfrac{x}{x+3} + \dfrac{1}{x-4} < 1$ _____

7. $\dfrac{x}{x+1} + \dfrac{2x}{x-1} > \dfrac{2}{x^2-1}$ _____

8. $\dfrac{2x}{x+2} - \dfrac{x}{x-3} < \dfrac{9}{x^2-x-6}$ _____

Technology
9.1 A Measure of Steepness

A quick glance at the graph of $y = \frac{1}{x}$ indicates that in the first quadrant the graph slants down to the right. Further examination of the graph suggests that the graph becomes straighter or flatter as x increases.

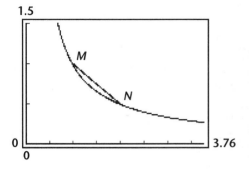

How can you devise a test to see if a graph becomes straighter or flatter?

One approach to the question is to draw $\overline{MN}$ with endpoints on the graph and x-coordinates that differ by 1. Then study the slope of $\overline{MN}$ as M slides along the curve to the right.

1. Write a formula for the slope of $\overline{MN}$ if M has coordinates $M(x, \frac{1}{x})$ and N has coordinates $N(x + 1, \frac{1}{x + 1})$

   _____

2. Create a spreadsheet in which column A contains x = 1, 2, 3, ... 8, 9, 10 and column B contains the slope of $\overline{MN}$.

3. How do the values in column B of the spreadsheet indicate that the graph of $y = \frac{1}{x}$ gets straighter or flatter as x increases?

   _____

Modify your spreadsheet to study the straightness of the graph of $y = \frac{k}{x}$ for each value of k.

4. $k = 2$ 5. $k = 3$ 6. $k = 4$

7. Use the approach above to study the steepness of $y = x^2$ for increasing values of x.

   _____

Technology
9.2 Linear Fractional Functions

Besides functions defined by reciprocals of linear expressions, there are functions defined by the quotient of two linear functions. Such functions can be called linear fractional functions and are defined by $y = \frac{ax + b}{cd + d}$.

Use a graphics calculator to graph each function over an interval that gives a complete picture of the graph.

1. $y = \frac{2x}{3x}$

2. $y = \frac{2x + 1}{3x - 2}$

3. $y = \frac{2x - 4}{3}$

4. $y = \frac{x + 1}{x - 2}$

5. $y = \frac{2x - 1}{3x}$

6. $y = \frac{x}{x}$

7. $y = \frac{2x + 2}{2x - 2}$

8. $y = \frac{x}{x - 1}$

9. Describe the graph of $y = \frac{ax + b}{cx + d}$, where $a = 0$, $c = 0$, but $b \neq 0$ and $d \neq 0$.

_____

10. Describe the graph of $y = \frac{ax + b}{cx + d}$, where neither $a \neq 0$, $d \neq 0$, but $c \neq 0$.

_____

11. Describe the graph of $y = \frac{ax + b}{cx + d}$, where neither a nor c are 0 but b and d are equal to 0.

_____

12. Describe the graph of $y = \frac{ax + b}{cx + d}$, where neither a, b, c, nor d are equal to 0.

_____

Technology
9.3 The Family $f(x) = \dfrac{1}{x^2 + c}$

In Lesson 9.3, you learned that graphing the reciprocal of a polynomial function results in a graph that is remarkably different from the graph of the function defined by the denominator.

To find more remarkable results, choose a particular family of functions for the denominator. Let $g_c(x) = x^2 + c$, where c is a fixed real number.

Use a graphics calculator to graph each function on the same calculator display.

1. $f(x) = \dfrac{1}{x^2 + 3}$

2. $f(x) = \dfrac{1}{x^2 + 2}$

3. $f(x) = \dfrac{1}{x^2 + 1}$

4. $f(x) = \dfrac{1}{x^2 - 1}$

5. $f(x) = \dfrac{1}{x^2 - 2}$

6. $f(x) = \dfrac{1}{x^2 - 3}$

7. Add the graph of $y = \dfrac{1}{x^2}$ to the graphs you constructed in Exercises 1–6.

8. Describe the graph of $f_c(x) = \dfrac{1}{x^2 + c}$ if $c > 0$.

_____

_____

9. Describe the graph of $f_c(x) = \dfrac{1}{x^2 + c}$ if $c = 0$.

_____

_____

10. Describe the graph of $f_c(x) = \dfrac{1}{x^2 + c}$ if $c < 0$.

_____

_____

Technology
9.4 Exploring Points That Slide Along Graphs

The diagram shows the graph of a quadratic function and right triangle OPN. Suppose that you let point P slide along the graph.

You can use rational functions to study the slope of $\overline{OP}$ as P moves along the graph of various quadratic functions.

$$\text{Slope of } \overline{OP} = \frac{PN}{ON} = \frac{y\text{-coordinate of P}}{x\text{-coordinate of P}}$$

If you are given an equation that tells the y-coordinate of P in terms of x, a spreadsheet will easily describe how the slope of $\overline{OP}$ changes as x changes.

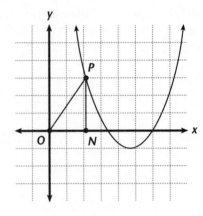

In Exercises 1 and 2, let $y = \dfrac{x^2 + 1}{x}$.

1. Create a spreadsheet in which column A contains $x = -4, -3.75, -3.50, ..., 3.50, 3.75,$ and 4.00 and column B contains the corresponding value of $\dfrac{x^2 + 1}{x}$. What trend do you see in the table of values?

   _____

2. Graph $y = \dfrac{x^2 + 1}{x}$ on a graphics calculator for values of x between -4 and 4 inclusive. Does the graph confirm your observations from Exercise 1?

   _____

In Exercises 3 and 4, let $y = \dfrac{x^2 - 1}{x}$.

3. Create a spreadsheet in which column A contains $x = -4, -3.75, -3.50, ..., 3.50, 3.75,$ and 4.00 and column B contains the corresponding value of $\dfrac{x^2 - 1}{x}$. What trend do you see in the table of values?

   _____

4. Graph $y = \dfrac{x^2 - 1}{x}$ on a graphics calculator for values of x between -4 and 4 inclusive. Does the graph confirm your observations from Exercise 3?

   _____

5. Suppose that $y = f(x)$ is a quadratic function and that your spreadsheet indicates that the slope of $\overline{OP}$ is 0 when $x = c$ and $x = d$. What does this information tell you about the roots of the equation $f(x) = 0$? (To help you answer the question, refer to Exercises 3 and 4.)

   _____

Technology
9.5 Checking Your Work with a Graphics Calculator

To find the sum of $\dfrac{3}{x+2}$ and $\dfrac{-2}{x+3}$, you would work as follows.

$$\dfrac{3}{x+2} + \dfrac{-2}{x+3} = \dfrac{3}{x+2} \cdot \dfrac{x+3}{x+3} + \dfrac{-2}{x+3} \cdot \dfrac{x+2}{x+2}$$

$$= \dfrac{3(x+3) + (-2)(x+2)}{(x+2)(x+3)}$$

$$= \dfrac{3x+9-2x-4}{(x+2)(x+3)}$$

$$= \dfrac{x+5}{(x+2)(x+3)}$$

$$= \dfrac{x+5}{x^2+5x+6}$$

If you graph $y = \dfrac{3}{x+2} + \dfrac{-2}{x+3}$ and $y = \dfrac{x+5}{x^2+5x+6}$ on the same coordinate plane, you should see that the graphs coincide.

Use pencil and paper to find each sum or difference. Then use a graphics calculator to check your answers.

1. $\dfrac{8}{3x-2} - \dfrac{2}{3x-1}$

2. $\dfrac{x+3}{x+1} + \dfrac{2}{x+1}$

3. $\dfrac{4}{3x+2} + \dfrac{-2}{2x-1}$

4. $\dfrac{-2x+1}{x+3} - \dfrac{2x}{2x-1}$

Use pencil and paper to solve each equation. Then use a graphics calculator to check your answers.

5. $\dfrac{8x}{3x-2} = -4$

6. $\dfrac{3x+1}{4x-2} = 7$

7. $\dfrac{x-1}{6} = \dfrac{4}{x+1}$

8. $\dfrac{x+1}{-6} = \dfrac{4}{x+1}$

Lesson Activity

9.1 Applying Inverse Variation

Edna buys a season pass to the Great Gorge Ski Resort. The cost of the pass is $350. Each time she uses the pass, a punch mark is made. The following table shows the relationship between the number of punch marks p, and the cost c, in dollars for each punch mark.

p	1	2	5	10	25
c	350	175	70	35	14

1. Use a graphics calculator to plot the points in the table.

2. Using the power regression feature, find the power function that best fits these data points.

   _____

3. Examine the differences between the actual values and the values obtained from the power function. Why is the power function the best fit curve?

   _____

4. Graph the regression equation. How many punch marks does Edna

   need so that the cost of each punch mark will be $5? _____

Suppose the ski resort has a promotional sale of a 20% discount on all season passes. Edna purchases a season pass at $350 decreased by 20%.

5. Complete the table.

p	1	2	5	10	25
c					

6. Using the power regression feature, find the power function that best fits these data points.

   _____

7. How does the cost of each ski trip at the normal fee compare with the discounted cost?

   _____

Lesson Activity
9.2 Rational Functions, Square Roots, and Cobwebs

The rational function $f(x) = 0.5\left(x + \frac{n}{x}\right)$, where n is a positive number, can be used to approximate square roots.

Use your graphics calculator to approximate $\sqrt{10}$.

1. Enter the expression $0.5\left(x + \frac{10}{x}\right)$ in the Y= screen. Choose any value for x as your first approximation of $\sqrt{10}$. Then find the y-value. Use this y-value as your second approximation of $\sqrt{10}$. Use the y-value as your new x-value and Zoom In to find the next y-value. Use this new y-value for your third approximation of $\sqrt{10}$ and find the new y-value. Repeat the process three more times. What is your final estimate of $\sqrt{10}$?

    _____

2. Try three different approximations of $\sqrt{10}$. Repeat the process. Do the different approximations yield different estimates of $\sqrt{10}$? Describe the estimates.

    _____

Square roots can be found by graphing f. To approximate $\sqrt{2}$, select Seq from the MODE menu. Then from the WINDOW FORMAT menu, select Web. Turn off all stat plots. On the Y= screen, enter $0.5(U_{n-1} + N/U_{n-1})$ for U_n. (U_{n-1} is on the keyboard.) Return to the home screen and store 2 to N. Set the window variables to U_nStart = 0.01, V_n = 0, nStart = 0, nMin = 0, nMax = 10, Xmin = -10, Xmax = 110, Xscl = 20, Ymin = -11, Ymax = 110, Yscl = 20. Press TRACE to display the graph. Then press the right arrow key to trace the cobweb.

3. Describe what you notice as you trace the cobweb. What is the approximate value of $\sqrt{2}$?

    _____

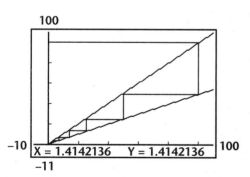

4. Approximate $\sqrt{17}$ and $\sqrt{30}$ by storing different values to N.

    _____

Square roots can be approximated by solving the equation $0.5x = \frac{0.5n}{x}$, where n is any positive number.

5. To approximate the $\sqrt{5}$, use your graphics calculator to graph $y = 0.5x$ and $y = \frac{0.5(5)}{x}$ on the same set of axes. Describe what the point(s) of intersection of the two graphs represent.

    _____

Lesson Activity
9.3 Magnifying Graphs

The accuracy and clarity of reciprocal polynomial function graphs drawn on graphics calculators depends on the resolution of the image on the screen. Viewing windows should be adjusted by changing scales so that the important regions of the graph can be viewed clearly.

For example, when $g(x) = \dfrac{1}{x^2 - 8}$ is graphed using a ZStandard window, the graph is difficult to view.

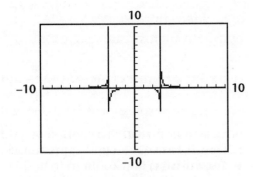

If you Zoom In around the origin, the graph almost disappears into the horizontal axis.

1. Graph $g(x)$ using the standard horizontal scale. Leave the horizontal scale unchanged and change the vertical scale as listed below.

   Vertical Scale: Ymin = −5, Ymax = 5; Ymin = −3, Ymax = 3; Ymin = −1, Ymax = 1

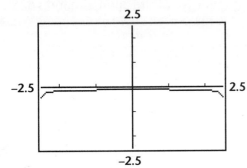

2. Compare the different views of $g(x)$ as the vertical scale is changed. Which scale gives the best view? _____

3. Graph $g(x) = \dfrac{1}{x^2 + 8}$ using several vertical scales. Which vertical scale gives the best view? _____

It is often necessary to change both the horizontal and vertical scales to graph a reciprocal function. For example, in ZStandard view the asymptotes of $g(x) = \dfrac{1}{(x^2 - 8)(x - 3)}$ are barely visible. If you Zoom In around the origin, part of the graph disappears into the horizontal axis.

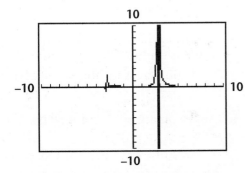

The function $g(x)$ is the reciprocal of the function $f(x) = (x^2 - 8)(x - 3)$. To get a good view of $g(x)$, it is necessary to find the zeros of f. Since the zeros of f are approximately −2.828, 2.828, and 3, the vertical asymptotes of g are at approximately $x = -2.828$, $x = 2.828$, and $x = 3$. To view the graph of g near the vertical asymptotes $x = 2.828$, and $x = 3$, change the horizontal scale to Xmin = 2.7, Xmax = 3.1, and Xscl = 0.6. Magnify the vertical scale to Ymin = −100, Ymax = 100, and Yscl = 50. The graph should be similar to the one shown.

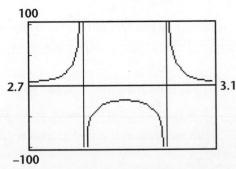

4. How would you adjust the scales to view g near the asymptote $x = -2.828$? _____

Lesson Activity

9.4 Can the Graph of a Function Cross Its Horizontal Asymptotes?

The horizontal asymptote for the function $f(x) = \dfrac{3x^2 + x - 2}{x^2 + 2x - 3}$ is $y = 3$.

To find the point where the function crosses the horizontal asymptote, substitute 3 for $f(x)$ and solve the equation for x. Here, $x = 1.4$ and the graph crosses its horizontal asymptote at $(1.4, 3)$.

Find the horizontal asymptote(s), then determine whether the rational function crosses its horizontal asymptote(s). If so, at what point(s)?

1. $f(x) = \dfrac{x^2 - 5x + 4}{x^2 - 2x - 3}$ _____

2. $f(x) = \dfrac{2(x + 2)(x - 1)}{(x - 2)(x + 1)}$ _____

3. $f(x) = \dfrac{x - 3}{x - 1}$ _____

4. $f(x) = \dfrac{x^2 - 9}{x^2 + 9}$ _____

5. For a rational function in which the degree of the numerator is equal to the degree of the denominator, does the function always cross the

horizontal asymptote? _____

Find the horizontal asymptote for each function. Then determine whether the rational function crosses its horizontal asymptote. If so, at what point(s)?

6. $f(x) = \dfrac{x + 3}{x^2 + 9}$ _____

7. $f(x) = \dfrac{x^2 - 3x + 2}{x^3 + 2x^2}$ _____

8. $f(x) = \dfrac{3x - 3}{x^2 - 9}$ _____

9. $f(x) = \dfrac{x^3 - 1}{x^4 - 1}$ _____

10. Does the rational function $f(x) = \dfrac{x - 2}{x^2 - 4}$ have a hole in it? Does the function cross its horizontal asymptote? Why or why not?

_____

11. Does the rational function cross the horizontal asymptote when the degree of the numerator is one degree less than that of the denominator? Explain.

_____

Lesson Activity
9.5 The Golden Ratio and Cobwebs

The golden ratio is found by dividing a segment into two parts so that the ratio of the length of the smaller part to the length of the larger part is equal to the ratio of the length of the larger part to the length of the entire segment. If the larger segment has a length of 1, the golden ratio can be defined as the solutions to the following equation.

$$\frac{x}{1} = \frac{1}{1 + x}$$

Simplifying gives the quadratic equation $x^2 + x - 1 = 0$. Using the quadratic formula, $x = \frac{-1 + \sqrt{5}}{2} \approx 0.618033989$ or $\frac{-1 - \sqrt{5}}{2} \approx -1.618033989$.

The golden ratio can be found graphically by approximating

$$\cfrac{1}{1 + \cfrac{1}{1 + \cfrac{1}{1 + \ldots}}}$$

To graph the golden ratio, select Seq from the MODE menu. From the WINDOW FORMAT menu, select Web. Turn off all stat plots. On the Y= menu, enter $1/(1 + U_{n-1})$ for U_n. (U_{n-1} is on the keyboard.) Set the WINDOW variables to U_nStart = 0.01, V = 0, nStart = 0, nMin = 0, nMax = 10, Xmin = 0, Xmax = 1, Xscl = 1, Ymin = 0, Ymax = 1, Yscl = 1. Press TRACE to display the graph. Then press the right arrow key to trace the cobweb.

1. Describe what you notice as you trace the cobweb.

    _____

2. Return to the home screen and store another value to U_{n-1}. Describe what you notice as you trace the cobweb.

    _____

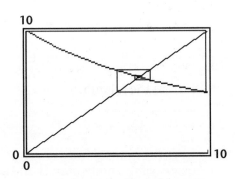

The reciprocal of the golden ratio can be approximated using the sequence:

$$\frac{1}{1}, \frac{1}{1}, \frac{2}{1}, \frac{3}{2}, \frac{5}{3}, \frac{8}{5}, \frac{13}{8}, \frac{21}{13}, \frac{34}{21}, \frac{55}{34}, \ldots$$

3. Use a calculator to find the value of the fifteenth term. Then find the value of the twentieth term, the thirtieth term. Describe what you notice as the terms increase.

    _____

Assessing Prior Knowledge
9.1 Inverse Variation

Find the constant of variation in each equation of direct variation.

1. $\frac{y}{3x} = 6$ _____

2. $\frac{x}{4y} = 5$ _____

- -

Quiz
9.1 Inverse Variation

1. Write an inverse variation equation with a constant of variation of 8. _____

In following the table, y varies inversely as the cube of x.

x	3	6	
y	2		0.2

2. Find the constant of variation. _____

3. Write the equation of variation. _____

4. Complete the table.

5. On a balanced seesaw, each person's distance from the fulcrum varies inversely as his or her weight. If an 80-lb child sits 5 ft from the fulcrum, how much does a child weigh who balances the seesaw sitting 4 ft from the fulcrum? _____

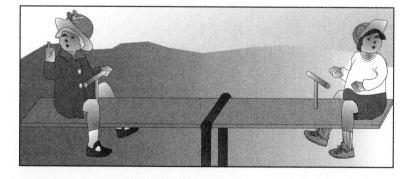

6. The time required to complete a landscaping job varies inversely as the number of workers. It takes 9 hours for 5 workers to complete the job. How long would it take 8 workers to complete the job? _____

Assessing Prior Knowledge
9.2 Simple Reciprocal Functions

Evaluate each fraction for the given values of *x*.

1. $\frac{1}{x + 3}$ when *x* is −7, −5, −3, −1, and 1 _____

2. $\frac{3x + 1}{2x - 3}$ when *x* is −1, 1, $\frac{3}{2}$, 2, and 4 _____

- -

Quiz
9.2 Simple Reciprocal Functions

Use your graphics calculator to graph each function. Identify the domain and range for each and find its asymptotes.

1. $f(x) = \frac{1}{x + 4}$ _____

2. $f(x) = \frac{x + 3}{x - 2}$ _____

The function $d(m) = \frac{100m}{m + 100}$, where *m* is the percent markup, is used by merchants to determine a discount percent that allows the merchant to sell an item at a predetermined price.

Find each of the following:

3. $d(20)$ _____   **4.** $d(30)$ _____

Water is mixed with 20 mL of a 30% acid solution.

5. Write the acid concentration as a function of the amount of water added.

_____

6. What is the concentration of the solution when 5 mL of water has been added?

_____

Assessing Prior Knowledge
9.3 Exploring Reciprocals of Polynomial Functions

Factor each expression.

1. $x^2 - x - 2$ _____

2. $2x^2 + 5x - 3$ _____

3. $2x^2 - 9x - 5$ _____

- -

Quiz
9.3 Exploring Reciprocals of Polynomial Functions

1. What is the domain of the function $f(x) = \dfrac{1}{x^2 + 5x + 6}$? _____

Determine whether each function is even, odd, or neither.

2. $f(x) = \dfrac{1}{x^2 + 3x - 4}$ _____

3. $f(x) = \dfrac{1}{x^3 + 4}$ _____

4. Construct a rational function that has a vertical asymptote at $x = 3$ and a horizontal asymptote at $y = 2$.

_____

Use a graphics calculator to graph each function, then describe its behavior.

5. $f(x) = \dfrac{1}{x^2 + 10x + 25}$ _____

6. $f(x) = \dfrac{1}{x^2 - 4x + 4}$ _____

7. $f(x) = \dfrac{1}{x^2 - 9}$ _____

8. $f(x) = \dfrac{1}{x^3 + 1}$ _____

Mid-Chapter Assessment
Chapter 9 (Lessons 9.1 – 9.3)

Write the letter that best answers the question or completes the statement.

_____ **1.** Given: y varies inversely as the square of x. If $y = 4$ when $x = 3$, the constant of variation is

   a. 12   **b.** 48   **c.** 36   **d.** $\frac{4}{9}$

_____ **2.** The function $f(x) = \frac{2x + 5}{x - 3}$ has a vertical asymptote at

   a. $y = 3$ **b.** $x = -3$ **c.** $y = 0$ **d.** $x = 3$

_____ **3.** If y varies inversely as the cube of x, and y is 3 when x is 4, the equation of variation is

   a. $x^3y = 192$ **b.** $xy^3 = 192$ **c.** $x^3y = 12$ **d.** $xy^3 = 12$

_____ **4.** The domain of the function $f(x) = \frac{1}{x^2 - 3x - 4}$ is

   a. all real numbers except $x = 4$ and $x = 1$
   b. all real numbers except $x = 4$ and $x = -1$
   c. all real numbers except $x = -4$ and $x = 1$
   d. all real numbers except $x = -4$ and $x = -1$

_____ **5.** The function $f(x) = \frac{x + 3}{x - 2}$ has a horizontal asymptote at

   a. $y = 2$ **b.** $x = 2$ **c.** $x = -3$ **d.** $y = 1$

6. Determine whether the function $f(x) = x^4 + x^2 + 3$ is odd, even, or neither. _____

7. Construct a rational function with a vertical asymptote at $x = -3$ and a horizontal asymptote at $y = 1$.

_____

8. Sketch the graph of the reciprocal of $f(x) = x^2 - 3x + 2$ on the grid provided.

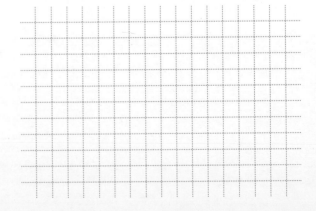

NAME _____ CLASS _____ DATE _____

Assessing Prior Knowledge
9.4 Quotients of Polynomial Functions

Divide each rational expression.

1. $\dfrac{2x^2 + 2x + 1}{x^2 - x - 2}$ _____

2. $\dfrac{4x^2 + 4x + 5}{2x^2 + 3}$ _____

- -

NAME _____ CLASS _____ DATE _____

Quiz
9.4 Quotients of Polynomial Functions

1. Identify the vertical asymptote(s) in $f(x) = \dfrac{(x - 2)(x + 5)}{x^2 - 4}$. _____

2. Identify the zeros of the rational function $f(x) = \dfrac{x^2 - 4}{x^2 - 25}$. _____

Simplify each rational function.

3. $f(x) = \dfrac{x^2 - 25}{x^2 - 7x + 10}$ _____

4. $f(x) = \dfrac{x^2 + 4x + 4}{(x + 2)^3}$ _____

Find the vertical and horizontal asymptotes for each function.
Give the domain of each.

5. $f(x) = \dfrac{3x}{x^2 + 2x + 1}$ _____

6. $f(x) = \dfrac{x(x + 2)(x - 4)}{(x - 2)(x + 3)(x + 4)}$ _____

7. Does the function $f(x) = \dfrac{x^2 - 16}{x + 4}$ have a hole in its graph? If so, at what
point? _____

Assessing Prior Knowledge
9.5 Solving Rational Equations

Add these fractions and simplify.

1. $\dfrac{2}{3} + \dfrac{5}{8}$ _____

2. $\dfrac{2}{x} + \dfrac{2x}{(x+1)}$ _____

- -

Quiz
9.5 Solving Rational Equations

Simplify each rational expression.

1. $\dfrac{3}{x+3} + \dfrac{x}{x-3}$ _____

2. $\dfrac{2}{x-4} + \dfrac{1}{x^2-16}$ _____

Use your graphics calculator to approximate all the solutions of the following equations.

3. $\dfrac{x^2 + 2x + 2}{x+2} = \dfrac{3}{2}$ _____

4. $\dfrac{x^2 + 2}{x^2 - 2} = \dfrac{4}{3}$ _____

5. $\dfrac{4x+3}{x^2 + x - 2} = 5$ _____

6. $\dfrac{8x+3}{x^2+3} = 1$ _____

Solve the rational equations.

7. $\dfrac{5}{x-3} = \dfrac{7}{x+2}$ _____

8. $\dfrac{3x+1}{3} + \dfrac{x+5}{2x+1} = \dfrac{7x+4}{6}$ _____

9. The reflection pool at the municipal park can be filled by a single pipe in 4 hours. If a second pipe is installed which can fill the pool in 5 hours, how long will it take to fill the pool with both pipes open? _____

Chapter Assessment
Chapter 9, Form A, page 1

Write the letter that best answers the question or completes the statement.

_____ **1.** If $f(x) = \dfrac{1}{x - 3}$, then the domain of the function is

a. all real numbers
c. all real numbers except $x = -3$

b. all real numbers except $x = 3$
d. all real numbers except $x = 0$

_____ **2.** Given the function $f(x) = \dfrac{x - 6}{x^2 - 36}$, vertical asymptote(s) exist at

a. $x = 6$ b. $x = -6$ c. $x = \pm 6$ d. $x = 36$

_____ **3.** Which is an odd function?

a. $f(x) = x + 1$
c. $f(x) = x^3 + x$

b. $f(x) = x^2 + 2x + 1$
d. $f(x) = x^4 + x^2 + 4$

_____ **4.** Which function's graph has a horizontal asymptote at $y = 2$?

a. $f(x) = \dfrac{2x + 5}{x - 3}$

b. $f(x) = \dfrac{3x^2 + 2x - 1}{x^2 + 5x + 2}$

c. $f(x) = \dfrac{3x + 2}{5}$

d. $f(x) = \dfrac{6x^3 + x^2 + x - 4}{2x^2 + 3x + 2}$

_____ **5.** If y varies inversely as the square of x and $y = 4$ when $x = 2$, the equation of variation is

a. $xy = 8$ b. $x^2y = 8$ c. $x^2y = 16$ d. $xy^2 = 16$

_____ **6.** If the area of a parallelogram is to be constant at 54, the base varies inversely as the altitude. The inverse variation equation is

a. $\frac{1}{2}bh = 54$ b. $b \div 2h = 54$ c. $b \div h = 54$ d. $bh = 54$

_____ **7.** The graph of $f(x) = -\dfrac{4}{x}$ is contained in quadrants

a. I, II   b. II, IV   c. I, III   d. III, IV

_____ **8.** If 20 ml of water is mixed with 50 ml of a 30% acid solution, the concentration of the new solution is about

a. 21% acid   b. 20% acid   c. 64% acid   d. 9% acid

Chapter Assessment
Chapter 9, Form A, page 2

_____ **9.** The zeros of $f(x) = x^3 + 3x^2 + 2x$ occur at

    a. $0, 1, 2$ **b.** $1, 3, 2$ **c.** $0, -1, -2$ **d.** $0, -2, 1$

_____ **10.** The vertical asymptotes of the reciprocal function of $f(x) = x^3 + 5x^2 + 4x$ are

    a. $x = 0, 5, 4$ **b.** $x = 1, 5, 4$ **c.** $x = 0, -1, -4$ **d.** $x = 0, -5, -4$

_____ **11.** The given graph represents

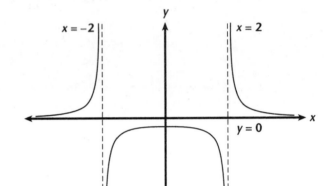

    a. $f(x) = \dfrac{x}{x^2 - 4}$

    b. $f(x) = \dfrac{1}{x^2 - 4}$

    c. $f(x) = \dfrac{1}{x^2 - 4x + 4}$

    d. $f(x) = \dfrac{x}{x^2 - 4x + 4}$

_____ **12.** 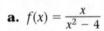$\dfrac{x + 5}{(x + 4)(x - 4)}$ is the simplified form of

    a. $\dfrac{2}{x + 4} + \dfrac{x + 3}{x - 4}$ **b.** $\dfrac{x}{x - 4} + \dfrac{5}{x + 4}$

    c. $\dfrac{x + 4}{x^2 - 16} + \dfrac{1}{x + 4}$ **d.** $\dfrac{1}{x - 4} + \dfrac{1}{x^2 - 16}$

_____ **13.** The solution of the rational equation $\dfrac{4}{x - 3} = \dfrac{6}{x + 2}$ is

    a. 3 **b.** 1 **c.** 13 **d.** -13

_____ **14.** The graph of the function $f(x) = \dfrac{x + 5}{x^2 - 25}$ has a hole at

    a. $x = -5$ **b.** $x = 5$ **c.** $x = \pm 5$ **d.** $x = 25$

_____ **15.** The simplified form of $\dfrac{x^3 + 5x^2 + 6}{x^2 - 4}$ is

    a. $\dfrac{x(x^2 + 5x + 6)}{x^2 - 4}$ **b.** $\dfrac{x(x + 3)}{x + 2}$

    c. $\dfrac{x(x + 3)}{x - 2}$ **d.** $\dfrac{x(x + 2)}{x - 2}$

_____ **16.** If a motorcycle driver travels 40 mi at m miles per hour and then 50 miles at $m + 10$ miles per hour, then the average speed $s(m)$ is given by

    a. $s(m) = \dfrac{40}{m} + \dfrac{50}{m + 10}$ **b.** $s(m) = \dfrac{9m^2 + 90m}{9m + 40}$

    c. $s(m) = \dfrac{8100m + 3600}{m^2 + 10m}$ **d.** $s(m) = \dfrac{90}{m^2 + 10m}$

Chapter Assessment
Chapter 9, Form B, page 1

1. Identify the domain and range of the function $f(x) = \dfrac{1}{x + 4}$.

    _____

2. Find the horizontal and vertical asymptotes of the function
    $f(x) = \dfrac{x^2 - 2x}{x^2 + 7x + 12}$. _____

3. Determine whether the function $f(x) = 4x^3 + 1$ is odd, even, or neither.

    _____

4. Construct a rational function with horizontal asymptote $y = 2$ and
    vertical asymptote $x = -2$.

    _____

5. If y varies inversely as the square of x, and $y = 5$ when $x = 2$, write the
    equation of variation.

    _____

6. If the area of a parallelogram is to be constant at 36, the base varies
    inversely as the altitude. Write the inverse variation equation.

    _____

7. Which quadrants contain the graph of $f(x) = \dfrac{2}{x}$? _____

8. If water is added to 20 ml of a 40% acid solution, write the acid
    concentration as a function of the amount of water added.

    _____

9. What is the axis of symmetry for the graph of an even function?

    _____

10. What are the zeros of the function $f(x) = (x + 3)(x - 1)$?

    _____

11. Given $f(x) = x^3 - 3x^2 + 2x$. What are the vertical asymptotes of its

    reciprocal function? _____

Chapter Assessment
Chapter 9, Form B, page 2

12. Graph the function $f(x) = \dfrac{1}{x^2 + 3x + 2}$ on the grid provided.

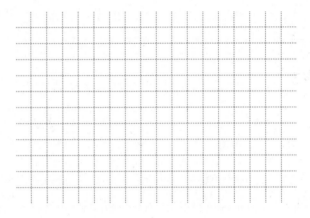

13. Sketch the graph of the reciprocal of $f(x) = x^3 + 4x^2 + 3x$ on the grid provided.

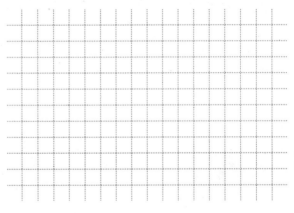

14. At what point does the function $f(x) = \dfrac{x + 3}{x^2 - 9}$ have a hole in it?

_____

15. Simplify the rational expression $\dfrac{x^2 + 5x + 4}{x^2 - 1}$. _____

16. Simplify the expression $\dfrac{5}{x^2} + \dfrac{x + 3}{x^3 - 5}$. _____

17. Use a graphics calculator to approximate all the solutions of the

equation $\dfrac{5x + 3}{x^2 + x + 2} = 2$. _____

18. Solve the rational equation $\dfrac{5}{x - 3} = \dfrac{-4}{2x + 6}$. _____

Alternative Assessment
Reciprocals of Polynomial Functions, Chapter 9, Form A

TASK: To graph reciprocals of polynomial functions

HOW YOU WILL BE SCORED: As you work through the task, your teacher will be looking for the following:

- whether or not you can identify the domain and range and find the vertical and horizontal asymptotes of a reciprocal function
- how well you can describe how to graph reciprocal functions as a transformation of the basic function $f(x) = \frac{1}{x}$

1. Explain how you can determine the horizontal and vertical asymptotes of $f(x) = \frac{1}{x + 2}$ without graphing.

   _____

   _____

2. Graph $f(x) = \frac{2x - 5}{x - 3}$ and $g(x) = 2 + \frac{1}{x - 3}$ on the same coordinate plane. Describe how the graphs are related. Explain how you can determine the horizontal and vertical asymptotes of f by using g.

   _____

   _____

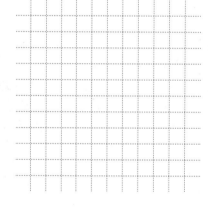

3. Graph the functions $p(x) = x^3 - 3x^2 - 9x + 27$ and $q(x) = \frac{1}{x^3 - 3x^2 - 9x + 27}$ on the grid provided.

4. What are the zeros of p and the vertical asymptotes of q? What is the domain of q?

   _____

   _____

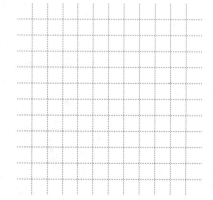

5. What are the turning points of p? How are the maximum and minimum points between the vertical asymptotes of q related to the turning points of p?

   _____

SELF-ASSESSMENT: Create a quadratic function. Find the inverse relation or inverse function. Find the reciprocal function. How are they alike? How are they different?

Alternative Assessment
Solving Rational Equations, Chapter 9, Form B

TASK: To solve rational equations

HOW YOU WILL BE SCORED: As you work through the task, your teacher will be looking for the following:

- how well you can solve rational equations by graphing
- whether you can solve rational equations by algebraic means

1. Describe how you would solve $\dfrac{x-3}{x} = \dfrac{x-4}{x-2}$ by graphing. Then find the solution.

   _____

   _____

2. Describe how you would solve $\dfrac{1}{x} + \dfrac{4}{x} + 1 = 6$ by algebraic means. Find the possible solutions. Explain why it is important to check the possible solutions.

   _____

   _____

3. Solve $\dfrac{2x+3}{x-1} - \dfrac{2x-3}{x+1} = \dfrac{10}{x^2-1}$ by algebraic means. Then solve the equation by graphing. Compare the two methods.

   _____

   _____

Let $f(x) = \dfrac{2x+1}{3x}$ and $g(x) = \dfrac{x-1}{3x}$.

4. Give the domains for f and g.

   _____

5. Find $f(x) + g(x)$ in simplified form. Then graph $h(x) = f(x) + g(x)$. Describe what you notice about the domain of h. Explain.

   _____

   _____

SELF-ASSESSMENT: Construct a rational function that has a hole in its graph at $x = 2$.

Practice & Apply

10.1 Parabolas

Determine the equation of each parabola.

1.

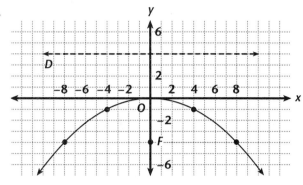

_____

2.

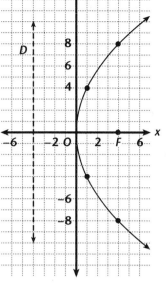

_____

Determine the vertex, axis of symmetry, focus, and directrix for each parabola.

3. $y = 3(x + 2)^2 - 1$ _____

4. $x = 2(y - 1)^2 + 3$ _____

When there is a change in the slope of a highway surface, such as at the top of a hill or the bottom of a dip, the roadway is often constructed in the shape of a parabola to provide a smooth transition between the two grades. Grade stakes are used to measure the offset or difference in elevation of the roadway with respect to the original grade.

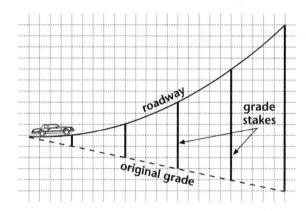

Suppose a parabolic-shaped roadway is modeled by the equation $y = 0.00025x^2$, where y is the offset and x is the horizontal distance to be covered.

5. If grade stakes are to be placed every 50 feet along the roadway, complete the table.

Grade Stake Distance	0	50	100	150	200
Offset					

6. Graph the equation on the grid provided.

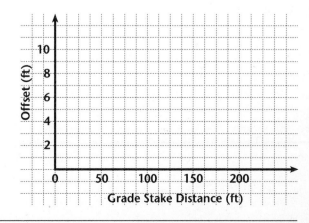

Practice & Apply
10.2 Circles

Write the equation in standard form for each circle.

1.

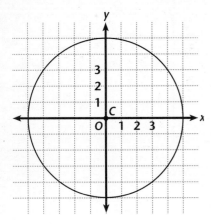

_____

2.

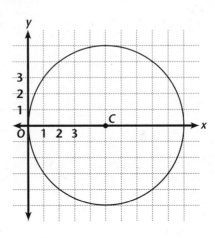

_____

3.

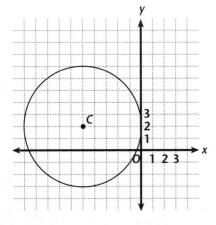

_____

4.

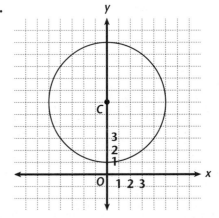

_____

Determine the center and radius of the circle represented by each equation.

5. $x^2 + y^2 = 4$

_____

6. $x^2 + 2x + y^2 + 10y + 22 = 0$

_____

7. $x^2 + y^2 + 10y = -21$

_____

8. $2x^2 + 2y^2 - 20y = -42$

_____

9. The circle with the equation $(x + 2)^2 + (y - 3)^2 = 25$ is translated 3 units to the right and 3 units downward. Find the equation of the

resulting circle. _____

Practice & Apply
10.3 Ellipses

Write the equation of each ellipse.

1.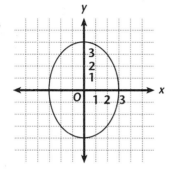

_____

2.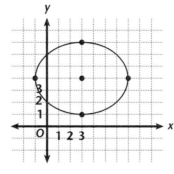

_____

3.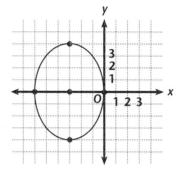

_____

Determine the vertices and co-vertices for each ellipse.

4. $x^2 + 4y^2 + 6x - 27 = 0$ _____

5. $4x^2 + y^2 - 8x - 6y - 23 = 0$ _____

Determine the coordinates of the foci for each ellipse.

6. $\dfrac{x^2}{25} + \dfrac{y^2}{9} = 1$ _____

7. $\dfrac{x^2}{36} + \dfrac{y^2}{100} = 1$ _____

The Earth's orbit is an ellipse with the Sun at one focus. The closest the Earth gets to the Sun is 147 million kilometers at perihelion. The Earth's farthest distance from the Sun is 152 million kilometers at aphelion.

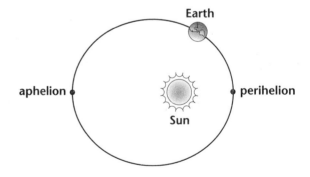

8. Write the equation of the elliptical path of the Earth around the Sun.

_____

9. Find the distance from the Sun to the other focus.

_____

10. A fireplace arch is constructed in the shape of a semi-ellipse. The opening is 2 feet high at the center and 5 feet wide along the base. Using (0, 0) as the center, write an equation that models the semi-elliptical fireplace.

_____

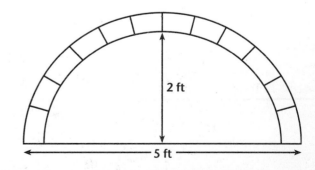

Practice & Apply
10.4 Hyperbolas

Write an equation for each hyperbola.

1.

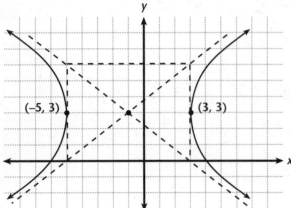

(−5, 3)   (3, 3)

_____

2.

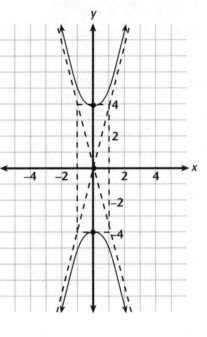

_____

Find the coordinates of the center, endpoints of the axes, and foci for each hyperbola.

3. $\dfrac{(x-1)^2}{4} - \dfrac{(y-4)^2}{25} = 1$ _____

4. $y^2 - \dfrac{x^2}{4} = 1$ _____

Write an equation for each of the following hyperbolas.

5. Transverse axis from (0, −3) to (0, 3); conjugate axis from (−4, 0)

to (4, 0). _____

6. Vertices at (2, −5) and (2, −7); foci at $(2, -6 + \sqrt{2})$ and $(2, -6 - \sqrt{2})$. _____

7. Foci at (−6, 0) and (6, 0) and a constant difference of 4. _____

8. Write the equation $y^2 - x^2 + 12y + 4x + 31 = 0$ in standard form.

Then find the coordinates of the center. _____

9. Tracking Station 2 is 300 miles east of
Station 1. Suppose a ship is sailing parallel
to the line connecting the stations. If the
time difference between the pulses
transmitted from the stations to the ship
is equivalent to a distance of 100 miles,
find the equation of the hyperbola on
which the ship could be located.

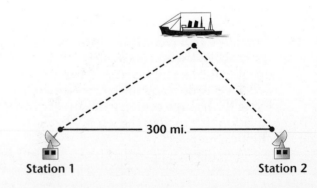

300 mi.

Station 1   Station 2

_____

Practice & Apply
10.5 Solving Non-linear Systems of Equations

**Solve each system by graphing on the grid provided.
Approximate roots to the nearest tenth where necessary.**

1. $9x^2 + 16y^2 = 144$
$3x + 4y = 12$

2. $\dfrac{x^2}{9} + \dfrac{(y-2)^2}{4} = 1$
$xy = 6$

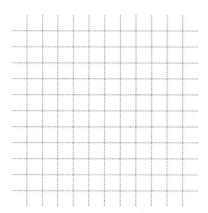
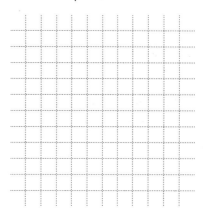

**Solve each system algebraically. Check using your graphics
calculator.**

3. $x^2 + y = 4$
$6x + y = 12$

4. $x^2 + y^2 = 16$
$x + y = 4$

5. $x^2 - 16y^2 = 16$
$x - 12y = -4$

_____

_____

_____

6. $4x^2 + y^2 = 20$
$x^2 + 4y^2 = 20$

7. $x^2 + y^2 = 16$
$2x^2 - y = -4$

8. $16x^2 + 9y^2 = 144$
$-48x + y^2 = 144$

_____

_____

_____

9. A rectangular plot of land is to be enclosed using two different kinds
of fencing. Two opposite sides will use heavy duty fencing selling for
$3 per foot and the remaining two sides will use standard fencing
selling for $2 per foot. What are the dimensions of the rectangular
plot if the area is 3750 square feet and the cost of the fencing is $600?

_____

10. A manufacturer has determined that the cost C (in thousands of
dollars) to produce x units of a product is given by the equation
$C = 100 + 25x$, and that the revenue R (in thousands of dollars) that
will realize a profit is given by the equation $R = 50x - x^2$. How many
units must the manufacturer produce and sell in order to break even
(that is, cost of production = revenue)?

_____

Practice & Apply
10.6 Exploring Parametric Representations of Conic Sections

Write a parametric representation for each equation.

1. $x^2 + y^2 = 25$

2. $\dfrac{x^2}{25} + \dfrac{y^2}{16} = 1$

3. $y = 4x^2 - x$

_____

_____

Write a rectangular equation for each system of parametric equations. Name the geometric object it represents.

4. $x(t) = 4\cos t$
$y(t) = 5\sin t$

5. $x(t) = 5t - 50t^2$
$y(t) = 5t$

6. $x(t) = 4\cos t$
$y(t) = 4\sin t$

_____

_____

Write a rectangular equation for each system of parametric equations. Name the geometric object represented. If the object is a circle, find the coordinates of the center and the length of the radius. If the object is an ellipse, find the coordinates of the center, and the vertices.

7. $x(t) = 3 + 4\cos t$
$y(t) = 5 + 4\sin t$

8. $x(t) = -3 + \cos t$
$y(t) = 5 + 4\sin t$

_____

_____

Write a parametric representation for each equation.

9. $x^2 + (y - 3)^2 = 25$

10. $\dfrac{x^2}{25} + (y + 3)^2 = 1$

_____

_____

11. $\dfrac{y^2}{25} = 1 - \dfrac{x^2}{9}$

12. $9x^2 + y^2 = 36$

_____

_____

Enrichment
10.1 Focusing In

When an object is thrown within the Earth's gravitational field, the object moves in a path described by a parabola.

1. What is the location of the highest point reached by a rocket fired from the ground called?

_____

2. A dart is thrown in a horizontal direction and follows a parabolic path. Suppose a dart thrower releases a dart 5 ft above the ground, and it hits the ground $10\sqrt{10}$ ft from the dart thrower. How high should a bulls-eye be placed at a distance of 10 ft from the thrower in order that the dart will hit it?

_____

3. A large spotlight is designed so that a cross section through its axis is a parabola and the light source is at the focus. Describe the location of the light source if the spotlight is 4 ft across at the opening and 2 ft deep?

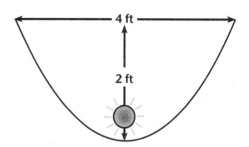

_____

4. Two 75-foot towers are 250 feet apart on a suspension bridge with a parabolic cable. Halfway between the towers, the vertex of the parabola is tangent to the road at the base of the towers. Find the height of the cable above the road at a point 50 feet from either tower.

_____

5. The vertical position of a bullet is given by $y = 16t^2$ and the horizontal position is given by $x = 40t$ for $t > 0$. By eliminating t between the two equations, show that the path of the bullet is a parabola.

_____

If a function can be described by a parabola that is concave up or concave down, it is easy to determine the maximum or minimum value of the function by using the vertex.

6. A homeowner plans to enclose a rectangular playing area for children. One side of the area will be formed by the wall of the house and the other sides will be formed by fencing. What is the maximum area that can be enclosed by 14 m of fencing?

_____

A farmer estimates that the profit (in dollars) from grazing x cattle on a pasture is given by the function $P(x) = -6.4x^2 + 320x - 3000$.

7. How many cattle should be grazed on this pasture to maximize profit? _____

8. What is the maximum profit? _____

9. Find the value of k for which the graph of $y = x^2 + kx + 25$ will have exactly one x-intercept.

_____

Enrichment
10.2 Rounding It Up

When an object is dropped into water, circular waves radiate away from the point of impact forming concentric circles.

Suppose you and a friend are standing on a straight section of the bank of a pond, and a rock is thrown into the water. If the speed of the waves is about 50 cm/s and the first wave will reach the bank where you are standing 50 seconds after the rock strikes the water, answer the following questions.

1. Using yourself as the origin, write an equation describing the possible locations of the rock striking the water.

   _____

2. What part of the equation can be eliminated immediately? _____

3. Suppose your friend is standing 10 meters to your right and the first wave reaches your friend 10 seconds after it reaches you. Write a second equation for the possible locations of the rock striking the water.

   _____

4. Use the grid to graph your equations from Exercises 1 and 3. Mark the possible point(s) of impact.

5. Which of these locations is most realistic? Explain your answer.

   _____

   _____

6. A fisherman uses the speed of these waves in water to judge the distance of his casts. If it takes 60 seconds for a wave to reach the fisherman on the shore, how far was the cast? _____

It takes sound roughly 5 seconds to travel one mile. Suppose you see a flash of lightning and hear the roll of thunder 15 seconds later.

7. What equation describes the possible locations of the lightning strike? _____

8. A friend who lives 4 miles north of you sees the same flash of lightning and hears the thunder 25 seconds after the flash. What equation describes the possible locations of the strike using your friend's data and your house as the origin?

   _____

Enrichment
10.3 Planetary Motions

The German astronomer Johannes Kepler discovered three laws of planetary motion related to ellipses. (One astronomical unit (AU) is the average distance between the Earth and the Sun.)

- The shape of each planet's orbit around the Sun is an ellipse with the Sun at one focus.

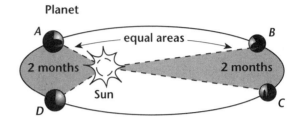

- An imaginary line drawn from a planet to the Sun sweeps out equal areas in equal intervals of time.

- The square of the time it takes a planet to orbit the Sun (in years) is equal to the cube of the planet's average distance from the Sun (in astronomical units.)

Use Kepler's laws to help answer Exercises 1–4.

1. How is a planet's average distance from the Sun related to the major axis of its elliptical orbit?

   _____

2. If the average distance between Mars and the Sun is 1.5 AU, how long does it take Mars to orbit the Sun?   _____

3. If an asteroid takes 3.2 years to orbit the Sun, on average, how far is it from the Sun in AUs?   _____

4. A satellite orbits the Earth in an elliptical path with the center of the Earth as one focus. If the satellite has a minimum altitude of 200 miles and a maximum altitude of 1000 miles above the surface of the Earth and the radius of the Earth is 4000 miles, what is the equation of the satellite's orbit?   _____

5. A semi-elliptical archway has a vertical major axis. The base of the arch is 10 feet across and the highest point of the arch is 15 feet above the ground. Find the height of the arch above a point on the ground 3 feet from the center.   _____

6. The focal width of an ellipse is the length of the line segment drawn perpendicular to the major axis through a focus with its endpoints on the ellipse. Find the focal width of the ellipse $\frac{x^2}{9} + \frac{y^2}{4} = 1.$ _____

Enrichment
10.4 Search and Locate

Because sound can travel through water, certain objects in the sea such as
submarines can be tracked by the sounds they make. Suppose two
microphones placed 10,000 meters apart underwater are designed to detect
sounds made by a submarine on the surface.

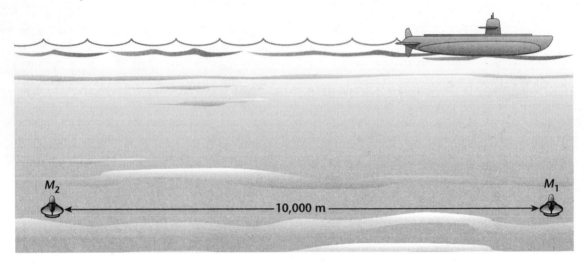

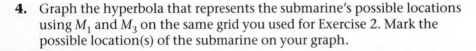

M_2 10,000 m M_1

1. Given that sound travels through water at
   the rate of 5000 m/s and a sound from the
   submarine reaches one microphone 0.8
   seconds after it reaches the other
   microphone, use a definition of a
   hyperbola and the distance formula to
   find the equation of the possible locations
   of the submarine.

   _____

2. Graph your equation on the grid provided.

3. Suppose a third microphone M_3 is located
   10,000 meters east of M_1. Using M_1 and M_3,
   a sound from the submarine reaches one microphone
   1 second after it reaches the other. How does this information
   help locate the submarine's position?

   _____

4. Graph the hyperbola that represents the submarine's possible locations
   using M_1 and M_3 on the same grid you used for Exercise 2. Mark the
   possible location(s) of the submarine on your graph.

5. What further information would you need to find the exact position of
   the submarine?

   _____

Enrichment
10.5 Conic Systems

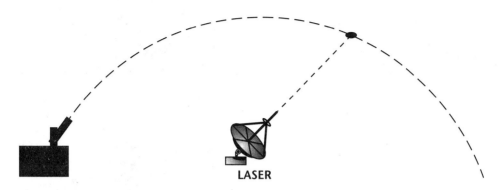

LASER

1. The drawing shows that a line and a parabola can intersect in one point. There are two other possible ways a line and parabola can intersect. Describe them.

   _____

2. There are three distinct ways a line can intersect a circle. Describe each possibility.

   _____

3. There are five possible ways a parabola and an ellipse can intersect. Describe the possibilities.

   _____

Solve each of the following problems by first forming a system of equations.

4. Steve made a model of two satellites orbiting the Earth. He represented his model on a coordinate grid. One satellite has a circular orbit of radius 5 about the origin. The second satellite has an elliptical orbit

   about the origin and passes through the points $(0, 4)$ and $\left(\frac{\pm 16\sqrt{7}}{7}, 0\right)$.

   Find the points at which the paths of the two satellites cross.   _____

5. Steve displayed his model on a rectangular stage. The stage has a perimeter of 42 m and an area of 54 m². Find the dimensions of the stage.   _____

6. Steve's workspace consists of a right triangle with squares along each leg. If the hypotenuse of the areas of the squares on each leg is 1 m², what is the length of each leg?

   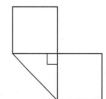

   _____

Enrichment
10.6 Search and Identify

Any conic section can be expressed in the form: $Ax^2 + Bxy + Cy^2 + Dx + Ey + F = 0$. If you use this form, you can easily identify what type of conic section an equation represents by calculating $B^2 - 4AC$, then looking on the chart.

$B^2 - 4AC$	Type of Conic
Negative	Ellipse
Zero	Parabola
Positive	Hyperbola

Identify each following conic.

1. $5x^2 - 7y^2 - 35 = 0$

_____

2. $x^2 + y^2 + 4x + 16y + 19 = 0$

_____

3. $x^2 - y - 3 = 0$

_____

4. $8x^2 + 6y^2 - 48 = 0$

_____

5. $xy - 1 = 0$

_____

6. $x^2 + y^2 + 10y + 16 = 0$

_____

7. $5x^2 + 7y^2 = 13$

_____

8. $y = -5x^2 - 7x + 13$

_____

9. $x^2 - 2x + y^2 - 2y + 1 = 0$

_____

10. $x^2 + y^2 - 10y = 0$

_____

11. $x^2 + 2x + 3y^2 - 5y + 3 = 5$

_____

12. $x^2 + 2x + y = 6$

_____

13. $x^2 + 3xy + 5y + 6 = 0$

_____

14. $4x^2 + 3xy + 3y^2 + 4y = 0$

_____

15. $x^2 + 3y^2 + y = 0$

_____

16. $3x^2 + 2xy + 4y^2 = 6$

_____

17. $3x^2 + 3xy - 3y^2 + 3y = 3$

_____

18. $2x^2 + 3xy - 4y^2 - 4y - 4 = 0$

_____

Technology

10.1 Problem Solving, Parabolas, and Area

The diagram shows a parabola with $\triangle ABC$ inscribed in it. Let $y = x^2$ and let A and B have coordinates $A(-2, 4)$ and $B(2, 4)$. Suppose that C has coordinates $C(x, x^2)$. To write an equation for the area of K of $\triangle ABC$, use the formula below.

$$K = \frac{1}{2} \times \text{base} \times \text{height}$$

$$= \frac{1}{2} \times AB \times CP$$

$$= \frac{1}{2} \times 4 \times |x^2 - 4|$$

$$K = 2\,|x^2 - 4|$$

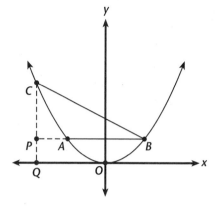

Use a graphics calculator as appropriate.

1. In the diagram, C is above $\overline{AB}$. Make a sketch of the inscribed triangle in which C is below $\overline{AB}$. Use both diagrams to explain how you know that the height CP of $\triangle ABC$ equal $|x^2 - 4|$.

    _____

2. Graph $K = 2|x^2 - 4|$. Use the graph to find all values of x (to the nearest tenth) for which $K = 1.5$ square units.

    _____

3. Use the graph from Exercise 2 to tell how many solutions $K = 2\,|x^2 - 4|$ has for a fixed value of K such that $0 < K < 8$, $K = 8$, and $K > 8$.

    _____

Refer to the diagram.

4. Write an equation for the area K of quadrilateral $ABDC$ given $A(-2, 4)$ and $B(2, 4)$.

    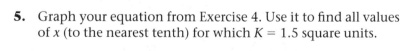

5. Graph your equation from Exercise 4. Use it to find all values of x (to the nearest tenth) for which $K = 1.5$ square units.

    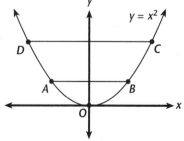

    _____

Technology

10.2 The Minimum Distance Between a Point and a Circle

Many interesting questions in coordinate geometry deal with questions of minimum distances. For example, look at the diagram. You have a point P inside a circle whose equation is known. Let Q be any point on the circle. Find the coordinates of Q such that PQ is minimized.

Suppose that P has coordinates $P(-1, 1)$ and the circle has equation $x^2 + y^2 = 9$. By following the strategy outlined below, you can find the minimum distance between P and the circle.

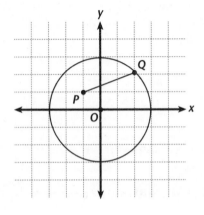

1. Using the equation for the circle, write the y-coordinate of Q in terms of x. (You should report two equations, one for the top half of the circle and one for the bottom half.)

   _____

2. Use the distance formula, the coordinates of P, and your answers to Exercise 1 to write equations for the distance d between P and Q.

   _____

3. Describe the range of x-coordinates for points on the circle. Give your answer as an inequality: XMIN $\leq x \leq$ XMAX.

   _____

4. Create a spreadsheet in which column A contains the smallest number in your answer to Exercise 3 and 0.1 increments of it. Column B contains the values of the expression for d from Exercise 2. From the spreadsheet, estimate the value(s) of x that gives the smallest value of d. Then find the corresponding y-coordinate of Q.

   _____

5. Let P have coordinates $P(4, 4)$. If q is on the circle with equation $x^2 + y^2 = 9$, find the minimum distance between P and Q.

   _____

Technology
10.3 Maximizing Area

The diagram shows the graph of $9x^2 + 16y^2 = 144$. Inscribed in the ellipse is rectangle $ABCD$. Below is a problem you can investigate with a graphics calculator.

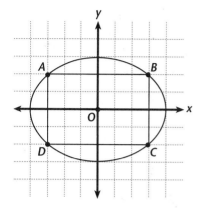

> Of all rectangles that you can inscribe in the graph of $9x^2 + 16y^2 = 144$, which one has the greatest or maximum area?

To solve the problem, notice that the area of rectangle $ABCD$ is given by $|2x| \, |2y| = 4|xy|$, where x and y are the coordinates of a point on the ellipse in the first quadrant.

Suppose that $9x^2 + 16y^2 = 144$.

1. Write an equation, in terms of x, for the area of rectangle $ABCD$ by using the equation Area = $4|xy|$.

   _____

2. Graph the equation you wrote in Exercise 1 over the interval $0 \leq x \leq 9.4$.

   _____

3. From the graph, find x (to the nearest tenth) that maximizes the area of rectangle $ABCD$ and the maximum area of $ABCD$.

   _____

Suppose that $4x^2 + 25y^2 = 100$.

4. Write an equation for the area of rectangle $ABCD$. Graph the equation you wrote over the interval $0 \leq x \leq 9.4$.

   _____

5. From the graph, find x (to the nearest tenth) that maximizes the area of rectangle $ABCD$ and the maximum area of $ABCD$.

   _____

6. Suppose that $ax^2 + by^2 = ab$. Make a conjecture about the maximum area of $ABCD$.

   _____

Technology
10.4 Stationary Vertices and Floating Foci

A hyperbola with its center at the origin and a horizontal transverse axis has an equation of the form $\frac{x^2}{a^2} - \frac{y^2}{b^2} = 1$. Once you choose values for a and b, you determine the hyperbola. Because $c^2 = a^2 + b^2$, you have also determined the foci. Alternatively, if you choose a and $c > a$, you have determined the hyperbola as well.

A graphics calculator is a useful tool for studying what happens if you keep the vertices $(a, 0)$ and $(-a, 0)$ stationary and let the foci $(c, 0)$ and $(-c, 0)$ float along the x-axis. To study the question, you will need to understand the statement shown.

If $\frac{x^2}{a^2} - \frac{y^2}{b^2} = 1$, then $\frac{x^2}{a^2} - \frac{y^2}{c^2 - a^2} = 1$ and $y = \pm \sqrt{(c^2 - a^2)\left(\frac{x^2}{a^2} - 1\right)}$.

Use the statement above as a positive and negative equation in your graphics calculator to graph each hyperbola for $a = 4$. Graph each value of c on the same calculator screen.

1. $c = 5$ **2.** $c = 6$ **3.** $c = 7$ **4.** $c = 8$

_____ _____ _____ _____

5. What appears to be the effect of keeping the vertices at $(4, 0)$ and $(-4, 0)$ but letting the foci float out to the right and the left of $(4, 0)$ and $(-4, 0)$, respectively?

_____

Use the statement above and a graphics calculator to graph each "hyperbola" for $a = 4$ and each value of c on the same calculator screen.

6. $c = 0$ **7.** $c = 1$ **8.** $c = 2$ **9.** $c = 3$

_____ _____ _____ _____

10. What appears to be the effect of keeping the vertices at $(4, 0)$ and $(-4, 0)$ but letting the foci float out to the left and the right of $(4, 0)$ and $(-4, 0)$, respectively?

_____

Technology
10.5 Solving Systems Using Matrices

In earlier lessons, you learned how to solve a system of linear equations by using the inverse of a matrix. You may be surprised to discover that you apply the same logic to solving certain systems of nonlinear equations. Examine the system of equations shown. Notice that each equation contains x^2 and y^2. Because of this, you can write the matrix equation found below the system.

$$\begin{cases} 2x^2 + 5y^2 = 10 \\ 3x^2 - 2y^2 = 6 \end{cases}$$

$$\begin{bmatrix} 2 & 5 \\ 3 & -2 \end{bmatrix} \begin{bmatrix} x^2 \\ y^2 \end{bmatrix} = \begin{bmatrix} 10 \\ 6 \end{bmatrix}$$

If you find the product $\begin{bmatrix} 2 & 5 \\ 3 & -2 \end{bmatrix}^{-1} \begin{bmatrix} 10 \\ 6 \end{bmatrix}$, you will find x^2 and y^2. To find x and y, all you need to do is take the square roots.

1. **a.** Find the inverse of $\begin{bmatrix} 2 & 5 \\ 3 & -2 \end{bmatrix}$.

    b. Find the product $\begin{bmatrix} 2 & 5 \\ 3 & -2 \end{bmatrix}^{-1} \begin{bmatrix} 10 \\ 6 \end{bmatrix}$.

    c. Use the product you found in part b to solve the system above.

    _____

Use a graphics calculator and matrices to solve each system.

2. $\begin{cases} x^2 + y^2 = 16 \\ 9x^2 + 25y^2 = 225 \end{cases}$

3. $\begin{cases} x^2 + y^2 = 16 \\ 9x^2 + 16y^2 = 144 \end{cases}$

4. $\begin{cases} x^2 + y^2 = 16 \\ 4x + 9y^2 = 36 \end{cases}$

5. $\begin{cases} x^2 + y^2 = 16 \\ 9x^2 - 16y^2 = 144 \end{cases}$

6. $\begin{cases} x^2 + y^2 = 16 \\ 4x^2 - 36y^2 = 144 \end{cases}$

7. $\begin{cases} x^2 + y^2 = 16 \\ 9x^2 - 9y^2 = 9 \end{cases}$

8. Use a graphics calculator to graph the system of equations in Exercise 2. Use the trace feature to find the solutions. Do your results confirm the solutions you found in Exercise 2?

    _____

9. Use a graphics calculator to graph the system of equations in Exercise 7. Use the trace feature to find the solutions. Do your results confirm the solutions you found in Exercise 7?

    _____

Technology
10.6 Exploring the Range for the Parameter *t*

You can begin to represent a circle with center at the origin and radius 3 by writing the parametric equations $x(t) = 3 \cos t$ and $y(t) = 3 \sin t$. However, these equations by themselves are not sufficient to describe the circle.

Use a graphics calculator to graph each pair of parametric equations over the specified range for *t*.

1. $x(t) = 3 \cos t$ and $y(t) = 3 \sin t$;

    $0 \le t \le \frac{\pi}{2} \approx 1.5708$

2. $x(t) = 3 \cos t$ and $y(t) = 3 \sin t$;

    $0 \le t \le \frac{3\pi}{2} \approx 4.7124$

3. Based on your results from Exercises 1 and 2, describe the smallest range for *t* that will give the complete circle.

   _____

Suppose that $x(t) = -2 + 3t$ and $y(t) = 1 + 2t$. Graph these equations over each specified interval for *t*.

4. $0 \le t \le 1$

5. $0 \le t \le 2$

6. $0 \le t \le 3$

   _____   _____   _____

7. What geometric shape will you get if you graph $x(t) = -2 + 3t$ and $y(t) = 1 + 2t$ over the interval $a \le x \le b$?

   _____

Suppose that $x(t) = 2t$ and $y(t) = t^2 + 1$. Graph these equations over each specified interval for *t*.

8. $-0.5 \le t \le 0.5$

9. $-1 \le t \le 1$

10. $-2 \le t \le 2$

    _____   _____   _____

11. What geometric shape will you get if you graph $x(t) = 2t$ and $y(t) = t^2 + 1$ over the interval $-a \le x \le a$, where $a > 0$?

    _____

Lesson Activity

10.1 Generating Parabolas

Draw a line on a sheet of paper. Place point F
2 inches above the line as shown.

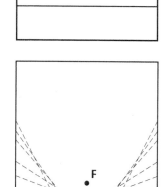

Hold the paper up to the light so you can see through
it. Fold the paper so that point F is on the line. Then
make a sharp crease. Continue to make different folds
with point F on the line and crease the paper at least
10 more times. The creases form tangents to a
parabola.

1. Examine the creases. Where are the vertex, focus,
and directrix of the parabola?

_____

2. Write an equation for the parabola in general
form.

_____

To create "conic graph paper," label a point
A on horizontally ruled paper. Then use a
compass to draw 10 concentric circles with
center A. Each time, increase the radius of
the circle so that the circle is tangent to a
pair of horizontal lines. Number the
concentric circles consecutively from 1 to
10, and number the lines starting with line
0 two units below point A, as shown.

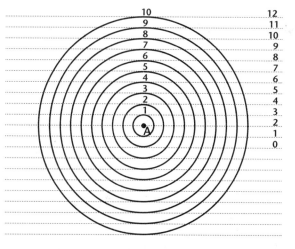

3. Plot the point where circle 1 intersects
line 1. Next plot the two points where
circle 2 intersects line 2. Repeat, plotting
the two points where each line intersects
each circle. Connect the points.

4. Explain why the curve is a parabola.

_____

_____

5. Where are the vertex, focus, and directrix of the parabola?

_____

_____

6. Draw a "family of parabolas" by using differently spaced horizontal
lines.

Lesson Activity
10.2 Circles and Cycloids

A cycloid is traced by the path of a point on the circumference of a circle rolling along a straight line.

From a sheet of paper, cut out a circle with radius 1 unit. Mark a point *P* on the circumference of the circle. On another sheet of paper, draw a number line. Place *P* at 0, then trace the path made by *P* as the circle is rolled along the line. When the circle rolls along the number line, the path traced by *P* is called a cycloid.

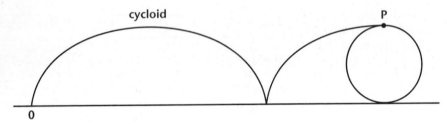

You can use a graphics calculator to trace point *P*'s path. With your calculator in parametric mode, set the parameter from 0 to 4π (approximately 12.5664) with a step size of $\frac{\pi}{30}$ (approximately 0.1047), *x*-values from 0 to 4π at increments of 1, and *y*-values from 0 to 2π (approximately 6.2832) at increments of 1.

1. Graph this system of parametric equations. What is the shape of the graph?

    $$\begin{cases} x(t) = t - \sin t \\ y(t) = 1 = \cos t \end{cases}$$

    _____

2. Use the trace feature to find the coordinates of point *P* when $t = 0$, 0.5π, π, 1.5π, 3π, and 4π.

    _____

3. Change the parameter using 0 to 8π (approximately 25.1327), and the *x*-values from 0 to 8π, then graph the cycloid. How many cycles are graphed?

    _____

4. What range values are needed to graph 8 cycles of the cycloid? Check by graphing this system of parametric equations.

    _____

Lesson Activity
10.3 Circles and Ellipses

Many interesting curves and patterns can be found by drawing sets of intersecting circles.

Make a sheet of conic graph paper. Start by drawing two points 8 units apart. Label the points *A* and *B*. Then use a compass to draw 12 concentric circles with center *A*. Increase the radius of each circle by one unit at a time. Then draw 12 concentric circles with center *B*, increasing the radius of each circle by one unit at a time. Label the circles as shown.

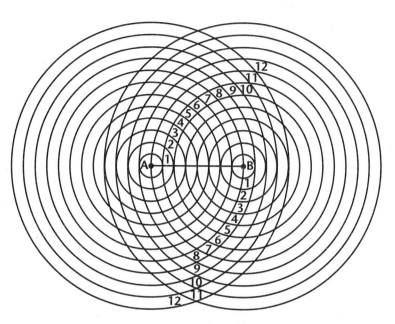

1. On your conic graph paper, plot all points where the sum of the distances is 10 units. For example, locate the point that is on the fourth circle from *A* and on the sixth circle from *B*. Connect the points with a smooth curve.

2. Explain why the curve is an ellipse. _____

3. Identify the center, foci, and vertices of the ellipse.

   _____

4. Write an equation of the ellipse in general form. _____

5. Draw a "family of ellipses" using different sums for the distance from point *A* and point *B*. Shade alternate regions to make a pattern.

   _____

6. Construct a circle on a sheet of paper. Place point *F* one unit from the center of the circle, as shown. Holding the paper up to the light, fold it so that point *F* falls on the circumference of the circle. Then make a sharp crease. Continue to fold and crease the paper at least 10 more times. What curve do the tangents form?

   _____

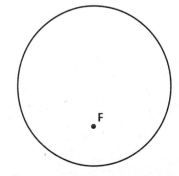

Lesson Activity
10.4 Hyperbolas and Curve Stitching

To curve stitch a hyperbola, start by drawing a circle with radius 2 units on a sheet on paper. Choose a point *F* about 1 unit outside the circle. Place a ruler on *F* so that the ruler intersects the circle in two points. Label these points of intersection *A* and *B*. Now place the ruler on *A* so that the ruler passes through the center of the circle, forming a diameter with *A* as one endpoint. Label the other endpoint of the diameter *C*. Next place the ruler on point *B* to find the endpoint of this diameter. Label the endpoint *D*. Draw ray *DA*. Then draw ray *BC*.

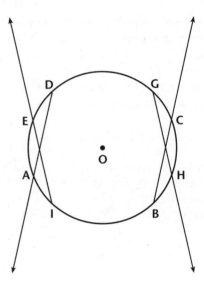

1. Continue to locate points on the circle, first locating points *E* and *G*, then forming diameters *EH* and *GI*, and drawing the corresponding rays.

2. Explain why the curve is a hyperbola.

_____

3. What is the constant difference? _____

4. Use the conic graph paper provided to generate a hyperbola, by plotting all points where the difference of the distances from *A* and *B* is 6 units.

5. Where are the center and foci of the hyperbola?

   _____

6. Write an equation of the hyperbola in general form.

   _____

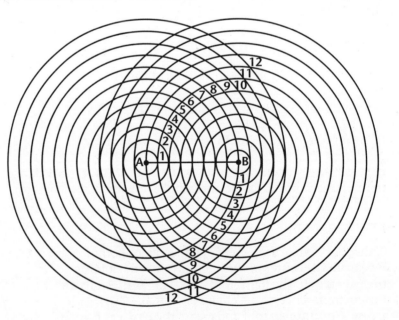

Lesson Activity
10.5 Graphing Circles and Polygons

The parameter T on the graphics calculator and the Tstep command are used to determined how frequently points are plotted when parametric equations are graphed. Varying the step value changes the shape of the graph.

With your graphics calculator in parametric mode, set the parameter from 0 to 2π (approximately 6.2832) with a step size of 0.5π (approximately 1.5707), x-values from -1.5 to 1.5 at increments of 1 and y-values from -1 to 1 at increments of 1.

1. Graph this system of parametric equations. What is the shape of the graph? $\begin{cases} x(t) = \sin t \\ y(t) = \cos t \end{cases}$

_____

2. Change the step size to 0.25π (approximately 0.7854). Graph the same system. What is the shape of the graph?

_____

3. Change the step size to 0.2π (approximately 0.6283). What is the shape of the graph?

_____

4. Determine the step size that will produce a graph of this system of parametric equations with the shape of a triangle.

_____

5. Change the step size to 0.4π (approximately 1.2566). Graph the system. What is the shape of the graph?

_____

6. What step sizes will produce a graph a septagon? A nonagon? _____

Varying the range values of the parameter T also changes the shape of the graph.

7. Change the step size to 0.8π (approximately 2.5132). What range is needed for T so that the same system of equations produces a 5-pointed star graph as shown?

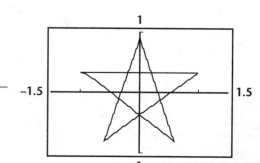

_____

8. Change the step size to $\frac{8\pi}{9}$ (approximately 2.7925). What range is needed for the parameter T, so that the system produces a 9-pointed star?

_____

Lesson Activity
10.6 Spirals

1. To draw a spiral, use a protractor to draw a series of lines 10 degrees apart about a common center on a sheet of paper. From the center, measure $\frac{1}{16}$ inch outward and mark this point on one of the lines. On the next line, mark the point $\frac{2}{16}$ inch from the center, mark $\frac{3}{16}$ inch on the next, and so on. Connect the points with a curved line.

With a graphics calculator in parametric mode, set the parameter from 0 to 4π (approximately 12.5664) with a step size of 0.1, x-values from -15 to 15 at increments of 1, and y-values from -10 to 10 at increments of 1.

2. Graph this system of parametric equations. What do you observe?

$$\begin{cases} x(t) = t\sin t \\ y(t) = t\cos t \end{cases}$$

_____

_____

3. Graph this system of parametric equations. What happens when the parameter is increased to 8π (approximately 25.1327)?

$$\begin{cases} x(t) = 0.5t\sin t \\ y(t) = 0.5t\cos t \end{cases}$$

_____

_____

4. Graph this system of parametric equations. Describe what happens to the graph as the coefficient of t decreases.

$$\begin{cases} x(t) = 0.1t\sin t \\ y(t) = 0.1t\cos t \end{cases}$$

_____

_____

5. Graph this system of parametric equations. Compare this system to the system graphed in Exercise 4.

$$\begin{cases} x(t) = 0.1t\cos t \\ y(t) = 0.1t\sin t \end{cases}$$

_____

_____

Various parameters and increments can make the spirals grow. Set the parameter from 0 to 500, with a step size of 0.1, x-values from -150 to 150 at increments of 10, and y-values from -100 to 100 at increments of 10.

6. Graph this system of parametric equations. What effect does t have on the graph when it is used as a coefficient?

$$\begin{cases} x(t) = t\cos t \\ y(t) = t\sin t \end{cases}$$

_____

_____

Assessing Prior Knowledge
10.1 Parabolas

Solve for y.

1. $\dfrac{3}{2a} = 5ay$ _____

2. $25a = 5ay$ _____

3. Write $y^2 + 10y + 25$ as the square of a binomial. _____

- -

Quiz
10.1 Parabolas

Determine an equation of the parabola with the following properties.

1. Focus at $(0, -3)$ and directrix $y = 4$. _____

2. Focus at $(2, 0)$ and vertex at $(0, 0)$. _____

Determine the vertex, focus, and directrix for the following parabolas. Identify the direction of opening.

3. $y^2 = 4x$ _____

4. $x^2 = 10y$ _____

5. $x^2 + 2x - 8y = -6$ _____

6. When a Little Leaguer throws a baseball upward, its path is a parabola that opens downward. If the ball reaches a height of 40 ft and lands 50 ft from where it was thrown, find the coordinates of the vertex and write an equation of the parabola.

_____

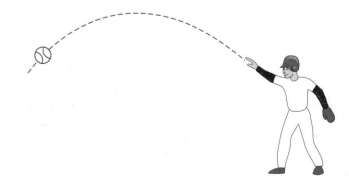

Assessing Prior Knowledge

10.2 Circles

1. Factor $y^2 + 12y + 36$. _____

Determine the distance between the following pairs of points.

2. $(1, 2)$ and $(3, 5)$ _____

3. $(1, 2)$ and (x, y) _____

- -

Quiz

10.2 Circles

Write the equation, in standard form, of the circle with center C and radius r.

1. $C(2, 4)$; $r = 5$ _____

2. $C(-6, 4)$; $r = \dfrac{5}{2}$ _____

3. $C\left(\dfrac{3}{5}, -\dfrac{1}{3}\right)$; $r = \sqrt{5}$ _____

4. $C(-4, -3)$; $r = 4$ _____

Determine the center C and radius r of the circle represented by each equation.

5. $x^2 + y^2 + 4x + 2y + 4 = 0$ _____

6. $x^2 - 4x + y^2 - 8y - 4 = 0$ _____

7. $3x^2 + 3y^2 + 12x - 6y = 18$ _____

8. $x^2 + y^2 + 6x - 8y - 7 = 0$ _____

9. The circle with equation $x^2 + y - 6x + 4y + 9 = 0$ is translated 3 units to the right and 3 units upward. Find the equation of the resulting circle.

_____

Assessing Prior Knowledge
10.3 Ellipses

Complete the square to solve each equation for *x*.

1. $x^2 = 5 - 4x$ _____

2. $3 = x^2 - 2x$ _____

3. $x^2 = 7 - 6x$ _____

4. $x^2 = 4x - 6$ _____

- -

Quiz
10.3 Ellipses

Write an equation of the ellipse for each of the following.

1. Vertices at $(\pm 3, 0)$ and $(0, \pm 5)$ _____

2. Vertices at $(\pm 8, 0)$ and $(0, \pm 4)$ _____

3. Vertices at $\left(\pm \frac{4}{3}, 0\right)$ and $\left(0, \pm \frac{9}{2}\right)$ _____

Determine the vertices and co-vertices for each ellipse.

4. $x^2 - 4x + 2y^2 + 6y + 2 = 0$ _____

5. $15x^2 + 3y^2 - 75x + 9y + 60 = 0$ _____

6. The orbit of a communications satellite about Earth is an ellipse with Earth at one focus. The closest the satellite gets to Earth is 5000 mi, while its farthest distance is 18,000 mi. Using (0, 0) as the center of the ellipse, write an equation for the orbit of the communications satellite.

_____

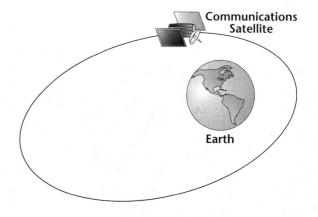

NAME _____ CLASS _____ DATE _____

Assessing Prior Knowledge
10.4 Hyperbolas

Write each equation of an ellipse in standard form.

1. $9x^2 + 16y^2 = 144$ _____

2. $49x^2 + 81y^2 = 3969$ _____

- -

NAME _____ CLASS _____ DATE _____

Quiz
10.4 Hyperbolas

Write an equation of the hyperbola for each of the following.

1. Transverse axis from (5, 0) to (−5, 0); conjugate axis from (0, −3) to (0, 3).

_____

2. Transverse axis from (0, −8) to (0, 8); conjugate axis from $(-\sqrt{11}, 0)$ to $(\sqrt{11}, 0)$.

_____

3. Foci at (−3, 0) and (5, 0) and a constant difference of 4.

_____

4. Vertices at (5, 3) and (1, 3); foci at (−1, 3) and (7, 3).

_____

Write each equation in standard form. Identify the coordinates of the center, lengths of the axes, and the coordinates of the foci.

5. $x^2 - y^2 + 10x + 6y + 7 = 0$ _____

6. $3x^2 - 3y^2 - 6x - 6y - 9 = 0$ _____

HRW material copyrighted under notice appearing earlier in this work.

Mid-Chapter Assessment
Chapter 10 (Lessons 10.1 – 10.4)

Write the letter that best answers the question or completes the statement.

_____ **1.** The equation $x^2 + y^2 - 3x + 4y + 6 = 0$ is best described as a(n)

     a. circle   **b.** ellipse   **c.** straight line   **d.** parabola

_____ **2.** The equation $x^2 + y^2 - 4x - 4y = 8$ is one of a circle with center and radius given.

     a. (0, 0); 4   **b.** (2, 2); 4   **c.** (2, 2); 8   **d.** (−2, −2); 4

_____ **3.** The equation of an ellipse with vertices at $(0, \pm 3)$ and $(\pm 6, 0)$ is

     a. $x^2 + 9y^2 = 35$ **b.** $x^2 + 4y^2 = 36$
     c. $4x^2 + y^2 = 36$ **d.** $6x^2 + 3y^2 = 36$

_____ **4.** A parabola whose equation is $x^2 + 4x - 4y + 12 = 0$ has focus and directrix

     a. $(-2, 3); y = 1$ **b.** $(2, 3); y = 3$
     c. $(-2, 2); y = 1$ **d.** $(2, 3); y = 1$

5. When a stone is thrown vertically upward, it tracks a path shaped like a parabola and reaches a height of 50 ft. The stone lands 30 ft from where it was thrown along a horizontal path. Write the equation of the parabola.

_____

6. The circle with equation $x^2 + y^2 - 2x + 6y - 6 = 0$ is translated 4 units to the left and 2 units downward. Find the equation of the resulting circle.

_____

7. Find the vertices and co-vertices of the ellipse whose equation is
$\dfrac{(x + 2)^2}{16} + \dfrac{(y - 3)^2}{9} = 1$.

_____

8. Find the vertex V and focus F of the parabola whose equation is
$y = \dfrac{1}{2}(x - 4)^2 + 3$.

_____

9. Find the coordinates of the center, the vertices, and the exact coordinates of the foci of the hyperbola $3y^2 + 12y - 4x^2 - 24x = 24$.

_____

Assessing Prior Knowledge
10.5 Solving Nonlinear Systems of Equations

Solve each equation for *a*.

1. $9ay = 45y$ _____

2. $\frac{3}{2}x = 6ax$ _____

3. $7ya = -63y$ _____

- -

Quiz
10.5 Solving Nonlinear Systems of Equations

Solve each system.

1. $\begin{cases} x^2 + y^2 = 13 \\ 2x + y = 4 \end{cases}$ _____

2. $\begin{cases} y = x^2 - 4 \\ y = 3x \end{cases}$ _____

3. $\begin{cases} 3x^2 + y^2 = 16 \\ x^2 + 2y^2 = 12 \end{cases}$ _____

4. $\begin{cases} y^2 - 4x^2 = 16 \\ 2x^2 + y^2 = 16 \end{cases}$ _____

5. $\begin{cases} x^2 + y^2 = 29 \\ xy = 10 \end{cases}$ _____

6. $\begin{cases} 3x^2 + 2y^2 = 21 \\ 2x^2 - 5y^2 = -24 \end{cases}$ _____

7. Find the dimensions of a rectangle whose perimeter is 60 in. and whose area is 224 in.^{2}.

_____

Assessing Prior Knowledge

10.6 Exploring Parametric Representations of Conic Sections

Complete each trigonometric identity.

1. $1 + \tan^2 \theta = $ _____

2. $\sin^2 \theta + $ _____ $= 1$

3. $\sec \theta = \dfrac{1}{}$

- -

Quiz

10.6 Exploring Parametric Representations of Conic Sections

Write a parametric representation for each of the following. Name the geometric object being represented.

1. $x^2 + y^2 = 16$ _____

2. $\dfrac{x^2}{16} + \dfrac{y^2}{25} = 1$ _____

3. $y = x^2 + 5$ _____

4. $(x - 3)^2 + (y - 2)^2 = 1$ _____

5. $\dfrac{(x - 2)^2}{25} + \dfrac{(y + 3)^2}{16} = 1$ _____

Find the rectangular equation for each parametric representation. Name the geometric object being represented.

6. $\begin{cases} x(t) = 7 \cos t \\ y(t) = 3 \sin t \end{cases}$ _____

7. $\begin{cases} x(t) = 8 \cos t \\ y(t) = 8 \sin t \end{cases}$ _____

Chapter Assessment
Chapter 10, Form A, page 1

Write the letter that best answers the question or completes the statement.

_____ **1.** An equation of a parabola with focus at $(4, -3)$ and directrix $x = -2$ is

     a. $x = \frac{1}{12}(y + 3)^2 - 1$ **b.** $y = \frac{1}{12}(x - 3)^2 + 12$

     c. $y = \frac{1}{12}(x + 3)^2 + 12$ **d.** $x = \frac{1}{12}(y + 3)^2 + 1$

_____ **2.** What is the vertex of a parabola whose equation is $x^2 = -3y$?

     a. $(3, -3)$ **b.** $(0, 0)$ **c.** $(-3, -3)$ **d.** $\left(1, -\frac{1}{3}\right)$

_____ **3.** The equation of a parabola with vertex at $(3, -1)$ and focus at $(3, 1)$ is

     a. $y = \frac{1}{3}(x - 8)^2 - 1$ **b.** $y = \frac{1}{8}(x + 3)^2 - 1$

     c. $y = \frac{1}{8}(x - 3)^2 - 1$ **d.** $y = \frac{1}{8}(x - 3)^2 + 1$

_____ **4.** The equation of the circle with center at $(4, -3)$ and radius 3 is

     a. $(x - 4)^2 + (y + 3)^2 = 3$ **b.** $(x + 4) + (y - 3) = 9$
     c. $(x - 4)^2 + (y + 3) = 9$ **d.** $(x + 4) + (y - 3)^2 = 3$

_____ **5.** The center and radius of the circle whose equation is $x^2 + y^2 + 6x - 2y - 4 = 0$ are

     a. $(3, 1); 14$ **b.** $(-3, 1); 14$
     c. $(-3, 1); \sqrt{14}$ **d.** $(3, -1); \sqrt{14}$

_____ **6.** The circle whose equation $(x + 2)^2 + (y - 3)^2 = 25$ is translated 4 units to the left and 1 unit downward. The equation of the resulting circle is

     a. $(x + 6)^2 + (y - 2)^2 = 25$ **b.** $(x - 2)^2 + (y - 4)^2 = 25$
     c. $(x - 6) + (y - 2)^2 = 20$ **d.** $(x + 2)^2 + (y - 3) = 20$

_____ **7.** The equation of the ellipse with vertices at $(\pm 8, 0)$ and $(0, \pm 4)$ is

     a. $\frac{x^2}{8} + \frac{y^2}{4} = 1$ **b.** $\frac{x^2}{64} + \frac{y^2}{16} = 1$
     c. $x^2 + 2y^2 = 48$ **d.** $2x^2 + y^2 = 8$

_____ **8.** The vertices for the ellipse whose equation is $3x^2 + 12x + 5y^2 + 10y + 2 = 0$ are

     a. $(-2, -1 \pm \sqrt{3})$ **b.** $(-2 \pm \sqrt{5}, -1)$
     c. $(-2, -1 \pm \sqrt{5})$ **d.** $(-2 \pm \sqrt{3}, -1)$

_____ **9.** The equation of the ellipse with foci at $(2, 0)$ and $(-2, 0)$ and whose constant sum is 6 is

     a. $\frac{x^2}{5} + \frac{y^2}{9} = 1$ **b.** $\frac{x^2}{9} + \frac{y^2}{7} = 1$

     c. $\frac{(x + 2)^2}{6} + \frac{(x - 2)^2}{6} = 1$ **d.** $\frac{x^2}{9} + \frac{y^2}{5} = 1$

Chapter Assessment
Chapter 10, Form A, page 2

_____ **10.** The equation of the hyperbola with transverse axis from (5, 0) to (−5, 0) and conjugate axis from (0, 7) to (0, −7) is

    a. $\dfrac{y^2}{49} - \dfrac{x^2}{25} = 1$ **b.** $\dfrac{x^2}{5} - \dfrac{y^2}{7} = 1$

    c. $\dfrac{x^2}{25} - \dfrac{y^2}{49} = 1$ **d.** $\dfrac{y^2}{5} - \dfrac{x^2}{7} = 1$

_____ **11.** The hyperbola whose equation is $3y^2 - 75x^2 = 225$ is expressed in standard form as

    a. $3y^2 = 225 + 75x^2$ **b.** $y - 25x^2 = 225$

    c. $y^2 = 75 + 25x^2$ **d.** $\dfrac{y^2}{75} - \dfrac{x^2}{3} = 1$

_____ **12.** The coordinates of the foci of the hyperbola whose equation is $16x^2 - 9y^2 = 9$ are

    a. $\left(\dfrac{5}{4}, 0\right), \left(-\dfrac{5}{4}, 0\right)$ **b.** $\left(0, \dfrac{9}{16}\right), \left(-\dfrac{9}{16}, 0\right)$

    c. $\left(\dfrac{25}{16}, 0\right), \left(-\dfrac{25}{16}, 0\right)$ **d.** $\left(0, \dfrac{5}{4}\right), \left(0, -\dfrac{5}{4}\right)$

_____ **13.** The dimensions of a rectangle whose perimeter is 42 m and whose area is 104 m2 are

    a. 2 m × 52 m   **b.** 13 m × 8 m   **c.** 26 m × 4 m   **d.** 21 m × 5 m

_____ **14.** The circle whose equation is $x^2 + y^2 = 36$ and the line whose equation is $x + y = 6$ intersect at

    a. no points   **b.** (6, 0), (0, 6)   **c.** (−6, 0), (0, 6)   **d.** (−6, 0), (0, −6)

_____ **15.** What are the solutions for the system of equations $x^2 + 4y^2 = 17$ and $3x^2 - y^2 = -1$?

    a. (1, 2), (−1, 2), (1, −2), (−1, −2)   **b.** (3, 8), (3, −8), (−3, 8), (−3, −8)

    c. (2, 1), (2, −1), (−2, 1), (−2, −1)   **d.** no solutions

_____ **16.** A parametric representation of $x^2 + y^2 = 25$ is

    a. $x(t) = \sin t + 5, y(t) = \cos t + 5$ **b.** $x(t) = \cos t + 5, y(t) = \sin t + 5$

    c. $x(t) = 25 \sin t, y(t) = 25 \cos t$ **d.** $x(t) = 5 \cos t, y(t) = 5 \sin t$

_____ **17.** The rectangular equation for the parametric representation $x(t) = 3 \cos t$ and $y(t) = 3 \sin t$ is

    a. $x^2 + y^2 = 3$ **b.** $3x^2 + 3y^2 = 1$

    c. $x^2 + y^2 = 9$ **d.** $(x - 3)^2 + (y - 3)^2 = 9$

_____ **18.** The geometric object represented by $x(t) = 4 \cos t - 1$ and $y(t) = 4 \sin t + 2$ is

    a. a circle   **b.** a hyperbola   **c.** an ellipse   **d.** a parabola

Chapter Assessment
Chapter 10, Form B, page 1

1. Write the equation of a parabola with focus at (0, 1) and directrix $y = -3$.

_____

2. Find the vertex of a parabola whose equation is $y = x^2 - 4x + 3$.

_____

3. Write the equation of a parabola whose vertex is at (2, 3) and whose focus is at (2, 5).

_____

4. Write the equation of a circle whose center is at (−3, 4) and whose radius is 5.

_____

5. Determine the center and radius of the circle whose equation is $(x - 4)^2 + (y + 3)^2 = 36$.

_____

6. If the circle whose equation is $(x - 1)^2 + (y + 3)^2 = 16$ is translated 4 units to the left and 2 units upward, write the equation of the resulting circle.

_____

7. Determine the equation of the ellipse having vertices at (±4, 0) and (0, ±8).

_____

8. Find the vertices of the ellipse whose equation is $25x^2 + 4y^2 = 100$.

_____

9. Write the equation of the ellipse with foci at (0, 2) and (0, −2) and with constant sum of 6.

_____

Chapter Assessment
Chapter 10, Form B, page 2

10. Find the dimensions of a rectangle whose perimeter is 84 cm and whose area is 432 cm^2.

_____

11. Write the equation of the hyperbola with transverse axis from (4, 0) to (−4, 0) and conjugate axis from (0, 8) to (0, −8).

_____

12. Express in standard form the hyperbola whose equation is $75y^2 − 3x^2 = 225$.

_____

13. Determine the coordinates of the foci of the hyperbola whose equation is $9y^2 − 16x^2 = 9$.

_____

14. Find the coordinates of the point(s) of intersection of the circle whose equation is $x^2 + y^2 = 25$ and the line whose equation is $x − y = −1$.

_____

15. Determine all solutions of the system of equations $x − y = −2$ and $4x^2 + 4x − y + 1 = 0$.

_____

16. Write a parametric representation of $x^2 + y^2 = 81$.

_____

17. Find the rectangular equation for the parametric representation $x(t) = 4 \cos t − 2$ and $y(t) = 4\sin t − 3$.

_____

18. Name the geometric object represented by the parametric system $x(t) = 2\cos t$ and $y(t) = 2 \sin t$.

_____

Alternative Assessment
Relationships, Chapter 10, Form A

TASK: To determine the relationship between the definition of a conic and the equation of a conic

HOW YOU WILL BE SCORED: As you work through the task, your teacher will be looking for the following:

- how well you can use the equation of a parabola to find the vertex, focus, and directrix for the parabola
- whether you can graph the equation of a circle using the standard equation of a circle
- how well you can write the equation of an ellipse in standard form and find the coordinates of the center, vertices, and the foci for the ellipse

1. Explain how to find the vertex, focus, and directrix for the parabola represented by $8y - x^2 + 6x = 25$. Then find the vertex, focus, and directrix.

   _____

   _____

2. Explain how you can graph the circle represented by the equation $x^2 + y^2 - 6x - 2y + 6 = 0$.

   _____

   _____

3. The equation $4x^2 + 3y^2 = 48$ represents an ellipse. Write the equation of the ellipse in standard form. Then find the coordinates of the center, vertices, and foci. How can you determine if the major axis is horizontal or vertical?

   _____

   _____

   _____

4. Write the equation of the ellipse that has vertices at $(5, 0)$ and $(-5, 0)$ and co-vertices at $(0, 3)$ and $(0, -3)$.

   _____

SELF-ASSESSMENT: Compare the types of symmetry determined by the graphs of a parabola, a circle, and an ellipse.

Alternative Assessment
Solving Non-Linear Systems of Equations, Chapter 10, Form B

TASK: To solve non-linear systems of equations

HOW YOU WILL BE SCORED: As you work through the task, your teacher will be looking for the following:

- how well you can use graphs to determine the number of real solutions of a non-linear system and to estimate the solutions
- whether you can use algebraic methods to find solutions of a non-linear system

1. In the space provided, draw all the possible intersections that can occur between a circle and an ellipse.

2. In the space provided, draw all the possible intersections that can occur between the graphs of one first-degree equation and one second-degree equation.

3. Describe how to solve the non-linear system shown by graphing. Then solve the system.
$$\begin{cases} y^2 - x^2 = 16 \\ x^2 + y^2 = 34 \end{cases}$$

_____

_____

4. Solve the non-linear system shown using algebra.
$$\begin{cases} x^2 - y^2 = 6 \\ xy = 4 \end{cases}$$

_____

SELF-ASSESSMENT: Write and graph a system of non-linear equations with only non-real solutions.

_____

ANSWERS

Practice & Apply — Chapter 7

Lesson 7.1

Year x	1990	1991	1992	1993
Ozone Level y	8.40	5.46	3.55	2.31

Year x	1994	1995	1996	1997
Ozone Level y	1.50	0.97	0.63	0.41

Year x	1998	1999	2000
Ozone Level y	0.27	0.17	0.11

2. 0.27   **3.** In the year 2000   **4.** 0.65

5. 7%   **6.** 1.07   **7.** 9674

8. $f(x) = 2500(1.07)^{(x)}$

9. $f(x) = 957{,}000{,}000{,}000(1.075)^{(x)}$

10. \$4065 billion

Lesson 7.2

1. $2^{\frac{1}{4}}, 3^{1.5}, 4^{\sqrt{2}}$ **2.** 0.038   **3.** 18.158

4. 245.105   **5.** 8.574

6. For each ordered pair, the y-coordinate of $g(x)$ is three times the y-coordinate of $f(x)$.

7. Each ordinate of $g(x) = a \cdot b^x$ will be a times the ordinate of $f(x) = b^x$.

8. $f(x) = 18500(1.0375)^x$

9. \$20,660   **10.** \$2568.75   **11.** \$2863.18

12. \$2569.46   **13.** \$2867.15   **14.** \$2569.62

15. \$2868.49   **16.** $10\frac{1}{2}$ h

Lesson 7.3

1. $\log_{25} 5 = \frac{1}{2}$ **2.** $\log 1000 = 3$

3. $\log_{0.25} 0.0625 = 2$ **4.** $\log_2 \frac{1}{4} = -2$

5. $2^5 = 32$ **6.** $10^{-3} = 0.001$ **7.** $5^0 = 1$

8. $4^2 = 16$ **9.** a   **10.** $x = 3$ **11.** $x = \frac{3}{2}$

12. no solution   **13.** $c = 5$ **14.** $x = 4$

15. $y = 3$

16.

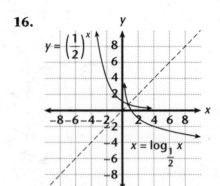

17. 1.44   **18.** 3.7

Lesson 7.4

1. $\log_{10} 1 - \log_{10} 3$

2. Possible answer: $\log_{10} 8 + \log_{10} 3$

3. $\log_5 2 + 4\log_5 a + \log_5 b + 3\log_5 c$

4. $2\log_2(y - 2) - \log_2(y + 1)$ **5.** 2   **6.** 10

7. 1   **8.** $\log_{10}\left(\frac{1}{3}\right)$ **9.** $\log_{10} 54$ **10.** $\log_{10} 25$

11. $\log_{10}\left(\frac{4}{9}\right)$ **12.** 0.778   **13.** 1.908

14. 0.239   **15.** -0.176 **16.** 42   **17.** 12

18. 2   **19.** $\sqrt{2}$

20.

Number of years after 1988	1	2	3	4
Interest rate (%)	10.32	7.53	5.90	4.74

Number of years after 1988	5	6	7	8
Interest rate (%)	3.85	3.11	2.49	1.95

Number of years after 1988	9	10
Interest rate (%)	1.48	1.06

ANSWERS

Lesson 7.5

1. 3.48   **2.** 0.48   **3.** 1.48   **4.** 2.48

5. 1.4   **6.** 10   **7.** 0   **8.** −0.60   **9.** 10

10. 0.01   **11.** 1.12

12. 1,000,000,000,000 or 1×10^{12}

13. 0.04   **14.** 117489.76   **15.** 1.05

16. 0.50   **17.** (4, 0); no y-intercept

18. Domain: $x > 3$; Range: all reals

19. The graph of f is translated 3 units to the right.

20. 1989   **21.** 1992   **22.** 2006   **23.** 6.0

24. 8.4

Lesson 7.6

1. 4.61   **2.** −3   **3.** 1.11   **4.** 1

5. $\ln 1 = 0$ **6.** $\ln 0.37 = -1$

7. $e^{2.69} = 14.73$ **8.** $e^{-0.69} = 0.5$

9. $x \approx 2.32$ **10.** $x \approx 1.55$ **11.** $x \approx 0.75$

12. 13,956   **13.** 37.6 years   **14.** 20.8 years

15. 30 words per min   **16.** 53 words per min

17. 74,082

18. The sales per unit decline levels off.

Lesson 7.7

1. $x = 5$ **2.** $x = 2$ **3.** $x = 12$ **4.** $x = 6$

5. $x = 1$ **6.** $x = \dfrac{e^3}{3}$

7. 9.27 billion or a loss of 9.27 billion

8. 1993   **9.** 2.13   **10.** 0.91   **11.** 1.15

12. 403.43   **13.** 0.92   **14.** 0.37

15. 1.001 volts

16. about 1 hour and 24 minutes

Enrichment — Chapter 7

Lesson 7.1

1. 921,577 people   **2.** 50%

3. 9.31×10^{-9}g   **4.** 22 yr   **5.** 11 yr

Lesson 7.2

1. $3278.99   **2.** $3706.86   **3.** $5373.23

4. $170.26   **5.** $4334.39   **6.** $7935.75

7. $6238.49   **8.** $5640.74   **9.** $7813.27

Lesson 7.3

1. $x = \dfrac{11}{7}$ **2.** $x = \dfrac{1 + \sqrt{5}}{2}$ **3.** $x = \dfrac{16}{3}$

4. $x = 2$ **5.** $x = \dfrac{13 + \sqrt{133}}{2}$ **6.** no solution

7. $x = \dfrac{3 + \sqrt{57}}{2}$ **8.** $x = \dfrac{19 + \sqrt{341}}{2}$

9. $x = 5$ **10.** $x = 2, 5$ **11.** no solution

12. $x = 4$

Lesson 7.4

1. $x = \dfrac{9}{4}$ **2.** $x = \dfrac{3 + \sqrt{5}}{2}, \dfrac{3 - \sqrt{5}}{2}$ **3.** $x = 8$

4. $x = 5$ **5.** $x = 6$ **6.** $x = 125$ **7.** $x = 8$

8. $x = 3$ **9.** $x = 4$ **10.** $x = 2\sqrt{2}$

11. $x = 6$ **12.** $x = \dfrac{1}{6}$

Lesson 7.5

1.

Number	Scientific Notation	Common Log	Log Expressed as Sum
3	3×10^0	0.477	$0 + 0.477$
30	3×10^1	1.477	$1 + 0.477$
300	3×10^2	2.477	$2 + 0.477$
3000	3×10^3	3.477	$3 + 0.477$
300,000	3×10^5	5.477	$5 + 0.477$

2. The decimal parts for each logarithm are the same.

3. The whole number part is equal to the exponent of the 10 when the number is written in scientific notation.

4.

Number	Characteristic	Mantissa
20,000	4	0.301
7,000,000	6	0.845
0.2	−1	0.301
0.07	−2	0.485
14	1	0.146
1.4	0	0.146

Lesson 7.6

1. 2   2. 2.25   3. 2.37037037

4. 2.44140625   5. 2.48832

6. 2.521626372   7. 2.546499697

8. 2.565784514   9. 2.581174792

10. 2.59374246   11. 2.704813829

12. 2.716923932   13. 2   14. 2.5

15. 2.666666667   16. 2.708333333

17. 2.716666667   18. 2.718055556

19. 2.718253968   20. 2.71827877

21. 2.718281526   22. 2.718281801

23. 2.718281826   24. 2.718281828

25. $1 + \frac{1}{1!} + \frac{1}{2!} + \ldots + \frac{1}{x!}$

26. $\left(1 + \frac{1}{x}\right)^x$; Less terms to enter into a calculator.

27. For the designated x value, add $\frac{1}{x!}$ to the result of the previous exercise which included all terms including $\frac{1}{(x-1)!}$

Lesson 7.7

1. 16 yr   2. 32.3%   3. 25 yr   4. $146,925

5. $1266   6. in the year 2003

Technology — Chapter 7

Lesson 7.1

1. 0.1   2. 5.72749995   3. 0.1

4. 2.2019×10^{19} 5. 0.1   6. 2.2019×10^{38}

7. As x increases, the difference increases as well.

8. As b increases, the 0.1 difference has more and more of an effect on the differences in column D.

Lesson 7.2

1. B4   2. B10   3. B6   4. not in table

5. B10   6. B6   7. $\sqrt{3^n}$ 8. $\sqrt{3^{11}}$, 3^(11/3)

9. To find $3^{7/5}$, calculate 3^(7/5).

Lesson 7.3

1. 0.20411998   2. 1.26007139

3. 2.12057393   4. 0.12057393

5. 3.26007139   6. 1.20411998

7. 3.12057393   8. 3.20411998

ANSWERS

9. Add n to log a.   **10.** 5.65513843

11. 2.57863921   **12.** 1.57863921

13. 1.96848295   **14.** 5.57863921

15. 0.65513843   **16.** −0.0315171

17. 2.65513843   **18.** Subtract n from log a.

Lesson 7.4

1.
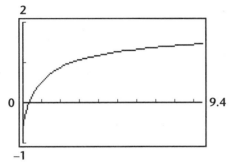

2.
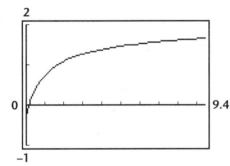

3.
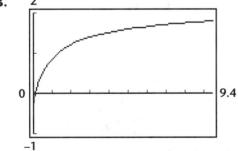

4.
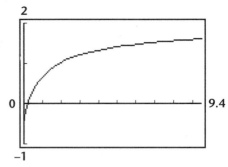

5. The pairs of graphs in Exercises 1–4 coincide. This suggests that the two expressions for y are equal.

6.
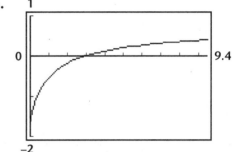

7.
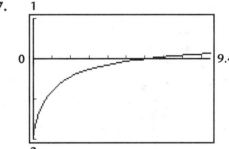

8.
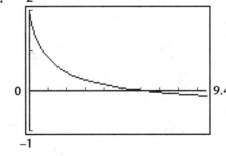

9.

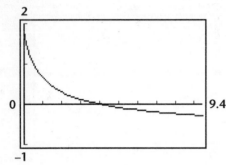

10. The pairs of graphs in Exercises 6–9 coincide. This suggests that the two expressions for *y* are equal.

Lesson 7.5

1. $\frac{x-1}{9}$ **2.** Check students' spreadsheets.

3. The differences are smaller near points *A* and *B*.

	A	B	C	D
1	1	0.0000	0.0000	0.0000
2	2	0.3010	0.1111	0.1899
3	3	0.4771	0.2222	0.2549
4	4	0.6021	0.3333	0.2687
5	5	0.6990	0.4444	0.2545
6	6	0.7782	0.5556	0.2226
7	7	0.8451	0.6667	0.1784
8	8	0.9031	0.8889	0.0654
9	9	0.9542	0.8889	0.0654
10	10	1.0000	1.0000	0.0000

4. $\frac{0.699}{4}(x-1)$ **5.** $\frac{0.301}{5}(x-10)+1$

6. Yes, the differences in column D are smaller and the approximation is improved.

	A	B	C	D
1	1	0.0000	0.0000	0.0000
2	2	0.3010	0.1748	0.1263
3	3	0.4771	0.3495	0.1276
4	4	0.6021	0.5243	0.0778
5	5	0.6990	0.6990	-3E-05
6	5	0.6990	0.6990	-3E-05
7	6	0.7782	0.7592	0.0190
8	7	0.8451	0.8194	0.0257
9	8	0.9031	0.8796	0.0235
10	9	0.9542	0.9398	0.0144
11	10	1.0000	1.0000	0.0000

7. It is reasonable to expect that if the differences in column D of the spreadsheet in Exercise 6 are correspondingly smaller than the differences in column D of the spreadsheet in Exercise 3, then the corresponding differences will be smaller when more points on the logarithmic curve and line segments are used to approximate log *x*. This will be true, of course, unless the *y*-coordinate of a point on one of the line segments is also on the logarithmic curve.

ANSWERS

Lesson 7.6

1.

	A	B	C
1		1.00000000	2.71828183
2	1	1.00000000	
3	2	0.50000000	
4	3	0.16666667	
5	4	0.04166667	
6	5	0.00833333	
7	6	0.00138889	
8	7	0.00019841	
9	8	2.4802E-05	
10	9	2.7557E-06	
11	10	2.7557E-07	
12	11	2.5052E-08	
13	12	2.0877E-09	

2.

	A	B
1	1	2
2	2	2.25
3	3	2.37037037
4	4	2.44140625
5	5	2.48832
6	6	2.52162637
7	7	2.5464997
8	8	2.56578451
9	9	2.58117479
10	10	2.59374246
11	11	2.60419901
12	12	2.661303529
13	13	2.62060089
14	14	2.62715156
15	15	2.63287872
16	16	2.6379285
17	17	2.64241438
18	18	2.64642582
19	19	2.65003433
20	20	2.65329771

The numbers get closer and closer to e. However, the table would have to be extended much more to get a close approximation.

3. 2.71828172

4. 2.71805556; The sum gives an approximation to e.

5. 19.1875, e^3

Lesson 7.7

1. $x = -1.9, x = 2.4$ **2.** $x = -0.8$ **3.** $x = 1$

4. no solution **5.** no solution **6.** $x = 0.8$

7. $x = 0.3$ **8.** $x = 1$ **9.** $x = 1.8$

10. If $m > 0$, there may be zero, one, or two solutions. If $m = 0$, there may be zero or one solution. If $m < 0$, there is one solution.

11. $x = 1.7$ **12.** $x = -0.4$ **13.** no solution

14. $x = 2.2$ **15.** $x = -3.1$

16. $x = -4.8, x = 0$

17. If $m > 0$, there is one solution. If $m = 0$, there may be zero or one solution. If $m < 0$, there may be zero, one, or two solutions.

Lesson Activities — Chapter 7

Lesson 7.1

1. $y = -982 + 932x$

2.

Hour (x)	0	1	2	3
Population (y)	100	200	400	800
Population (y_p)	−982	−50	882	1814
Residual ($y-y_p$)	1082	250	−482	−1014

Hour (x)	4	5	6
Population (y)	1600	3200	6400
Population (y_p)	2746	3679	4611
Residual ($y-y_p$)	−1146	−479	1789

ANSWERS

3. Except for $y = 200$, the residuals are either much larger or much smaller than the actual y-values.

4. The residuals at 0, 3, 4, and 6 hours.

5. There is no apparent trend in residuals.

6. $y = 100(2)^x$

7. The residuals are all 0.

8. The exponential model.

9. The exponential model.

Lesson 7.2

1. $5425.08   **2.** $5473.41

3. The amount increases.   **4.** $4989.18

5. The amount is lowered.

Lesson 7.3

1. $y = 13146$; $1350x$; differences: 104, −268.5, 482.93, −370.84, −155.11, 208.08

2. $y = 13524(0.8679)^x$; differences: −274, −209.98, 738.97, −83.1, −82.44, 208.08

3. Possible answer: difficult to determine

4.

x	0	1	2	3
log y	4.1222	4.0617	4.0386	3.9408

x	4	5
log y	3.8803	3.8198

5.

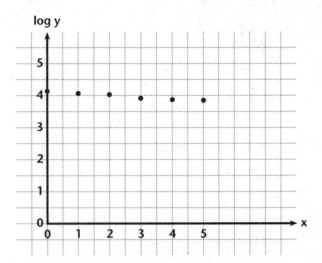

The graph is linear.

6. $r = -0.99145$ **7.** The exponential model.

Lesson 7.4

1.

0.6	0.9	1.2	1.5
2.853	2.7795	2.706	2.6325

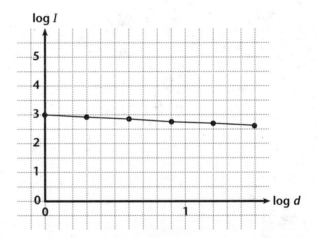

3. The graph is nearly linear.

4. The slope is −0.245 and the y-intercept is 3.

5. $y = 10^3 x^{-0.245}$

6.

log x	−0.3010	0	0.1761
log y	0.3802	−0.2218	0.5735

log x	0.3010	0.3979	0.4771
log y	−0.8239	−1.0177	−1.1739

ANSWERS

7.

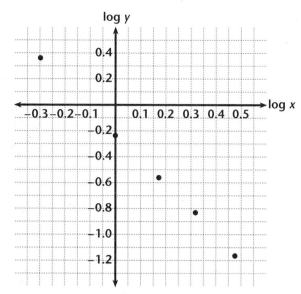

The graph is linear.

8. The slope is -2.0, the y-intercept is -0.22.

9. The data points (log x, log y) form a line.

10. $y = 0.6x^{-2}$

Lesson 7.5

1.

pH	0	1	2	3	4	5
[H⁺]	10^0	10^{-1}	10^{-2}	10^{-3}	10^{-4}	10^{-5}

pH	6	7	8	9	10	11
[H⁺]	10^{-6}	10^{-7}	10^{-8}	10^{-9}	10^{-10}	10^{-11}

2. Possible answer: Acid rain is rain water formed when neutral rain water combines with certain chemical pollutants in the atmosphere to form a slightly acidic solution as the rain water falls to the ground. Acid rain can lower the pH level of lakes.

3. Check students' responses. Dehydration: decreased; starvation: decreased; intoxication: increased; diabetic acidosis: decreased; diabetes insipidus: normal.

Lesson 7.6

1.

Partial Sum	Cumulative Sum
0.6166666667	0.6166666667
0.0365440115	0.6532106782
0.012929146	0.6661398242
0.0066076753	0.6727474995

2. The partial sums decrease.

3. It takes too long for the cumulative sum to converge to ln 2.

4.

Partial Sum	Cumulative Sum
0.653210678	0.653210678
0.0195368213	0.6727474993
0.0067036191	0.6794511184

5. The partial sums decrease.

6. This method is faster.

Lesson 7.7

1. $x = -2$

2. The flow rate is about 22.84 ft³/s.

3. The water level is approximately 4.5 ft.

Assessment — Chapter 7

Assessing Prior Knowledge 7.1

1. $6x - 3xy$ **2.** $4x(2y + x - 4y^2)$

Quiz 7.1

1. 1.007   **2.** 0.9497   **3.** 1.08   **4.** 6.25%

5. 1.0625   **6.** $6496.68   **7.** ~ 12 years

8. $f(t) = 4000(1.0625)^t$

Assessing Prior Knowledge 7.2

1. $a^{\frac{3}{10}}$ **2.** $\sqrt[10]{a^3}$

ANSWERS

Quiz 7.2

1. polynomial   2. exponential

3. polynomial   4. 11.66   5. 78.61

6.

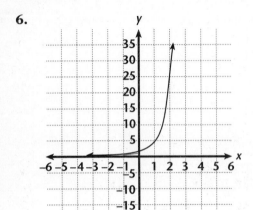

7. Translate or move each point on the graph of $f(x) = 5^x$ down 2 units.

8. $3133.99

9. Yes, because $4^{-x} = (4^{-1})^x = \left(\frac{1}{4}\right)^x$.

Assessing Prior Knowledge 7.3

1. $f^{-1}(x) = +\sqrt{x}$ or $f^{-1}(x) = -\sqrt{x}$

2. $f^{-1}(x) = \frac{1}{2}(x - 3)$

Quiz 7.3

1. $\log_5 \frac{1}{25} = -2$ 2. $\log_4 64 = 3$ 3. $\log_{36} 6 = \frac{1}{2}$

4. $4^2 = 16$ 5. $3^{-2} = \frac{1}{9}$ 6. $10^4 = 10,000$

7.

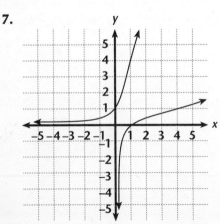

8. $\frac{5}{3}$ 9. 3

Assessing Prior Knowledge 7.4

1. 2^7 2. 5^3 3. 3^6

Quiz 7.4

1. $\log_4 3 + \log_4 5$ 2. $\log_{10} x + \log_{10}(x + 5)$

3. $\log_6 x + 3\log_6 y + 2\log_6 z$

4. $\log_2 25 - \log_2 16$

5. $\log_3 vw - \log_3 xy$ or $(\log_3 v + \log_3 w) - (\log_3 x + \log_3 y)$

6. $4\log_{10} x - 3\log_{10} y$ 7. 9   8. $\log_3 32$

9. $\log_2 6$ 10. ≈ 1.322

Mid-Chapter Assessment

1. c   2. b   3. d   4. b   5. 4.094

6. $p = \frac{\log_{10} C}{15}$ 7. $5157.42

8. 0.9598; population is decreasing.

Assessing Prior Knowledge 7.5

1. 1, 10, 100   2. $\frac{1}{10}, \frac{1}{100}$

Quiz 7.5

1. 2.57   2. -0.61 3. 3.58   4. -0.65

5. 398.11   6. 1.05   7. 0.15

8. $D = x > -3$
   R = all real numbers

9. All points on f are translated to the left 3 units compared with the points of $y = \log x$.

Assessing Prior Knowledge 7.6

1. Rational, since it repeats $1.333... = 1\frac{1}{3} = \frac{4}{3}$

ANSWERS

Quiz 7.6

1. 1.66 **2.** 0.09 **3.** 5.65

4. $\ln 0.67032 = -0.4$ **5.** $e^{1.0986} = 3$

6. 1.58 **7.** 0.74 **8.** $(3, 0)$

9. Graph of f is "stretched" horizontally; corresponding values of x are 3 times as far to the right as those of g.

Assessing Prior Knowledge 7.7

1. 4 **2.** 2 **3.** 7.4 **4.** 2

Quiz 7.7

1. $x = 3$ **2.** $x = 125$ **3.** $x = 5$

4. $x = 3.46$ **5.** $x = 1.13$

6. $x = 63,095,734.5$ **7.** $x = 49.5$

8. $x = 1.10$ **9.** $x = 12,088.38$ **10.** 6.75

Chapter 7 Assessment, Form A

1. b **2.** a **3.** b **4.** c **5.** b **6.** c **7.** d

8. c **9.** a **10.** b **11.** d **12.** b **13.** c

14. c **15.** c **16.** a **17.** c **18.** b

Chapter 7 Assessment, Form B

1. 0.986 **2.** 679,655 **3.** 4.146

4. 5180.31 **5.** $\log_3 \frac{1}{81} = -4$

6. $x = 2$ **7.** -2.57 **8.** 0.06 **9.** 2.71

10. $\ln 2.12 = 0.75$ **11.** $x = -5$

12. $x = 9.41$ **13.** about 28 years

14. $21,798.63 **15.** 1.6813

16. Graph of $g(x)$ is 1 unit below graph of $f(x)$ for all corresponding points.

17. $10^{1.4472} = 28$ **18.** 43,740

Alternative Assessment — Chapter 7

Form A

1.

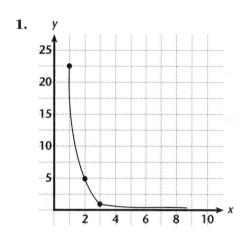

2. The domain is all real numbers and the range is all real numbers except $y \leq 0$.

3. f is an exponential function because the variable x is an exponent, and the base $\frac{1}{5}$, is positive and is not equal to 1.

4. Since the base of f, $\frac{1}{5}$, is between 0 and 1, f is a decay function.

5. The graph of g is a reflection through the y-axis of the graph of f.

6. f and g are inverse functions.

Score Point 4: Distinguished

The student demonstrates a comprehensive understanding of exponential and logarithmic functions. The student uses perceptive, creative, and complex mathematical reasoning throughout the task. He or she is able to use sophisticated, precise, and appropriate mathematical language throughout the task. Theoretical knowledge is apparent and applied to concrete situations as the student successfully demonstrates a comprehensive understanding of core concepts throughout the task.

ANSWERS

Score Point 3: Proficient

The student demonstrates a broad understanding of exponential and logarithmic functions. The student uses perceptive mathematical reasoning most of the time. He or she is able to use precise and appropriate mathematical language most of the time. Theoretical knowledge is apparent and applied to concrete situations as the student attempts to draw conclusions based on his or her investigations.

Score Point 2: Apprentice

The student demonstrates an understanding of exponential and logarithmic functions. He or she uses mathematical reasoning at times during the task. He or she uses appropriate mathematical language some of the time. Student attempts to apply theoretical knowledge to the task but may be able to draw conclusions from his or her investigation.

Score Point 1: Novice

The student demonstrates a basic understanding of exponential and logarithmic functions. He or she uses appropriate mathematical language some of the time. Theoretical knowledge is extremely weak and many responses are irrelevant or illogical. He or she may fail to follow directions and has great difficulty in communicating his or her responses.

Score Point 0: Unsatisfactory

Student fails to make an attempt to complete the task and his or her responses are just an attempt to fill the page or restate the problem.

Form B

1. Graph $y = \log(x + 9) - \log x$. Use the trace feature to find the x-value when y is 1. The solution is $x = 1$.

2. $x = 4$. Check the solution by substituting each value for x in the exponential equation.

3. $x = 2$ 4. $x = 3$ 5. $x = 2$

6. The graph shows no solution. By using the exponential-log properties a possible solution $x = -5$ is obtained. Since the log is not defined at $x = -5$, there is no solution.

Score Point 4: Distinguished

The student demonstrates a comprehensive understanding of solving exponential and logarithmic equations. The student uses perceptive, creative, and complex mathematical reasoning throughout the task. He or she is able to use sophisticated precise, and appropriate mathematical language throughout the task. Theoretical knowledge is apparent and applied to concrete situations as the student successfully demonstrates a comprehensive understanding of core concepts throughout the task.

Score Point 3: Proficient

The student demonstrates a broad understanding of solving exponential and logarithmic equations. The student uses perceptive mathematical reasoning most of the time. He or she is able to use precise and appropriate mathematical language most of the time. Theoretical knowledge is apparent and applied to concrete situations as the student attempts to draw conclusions based on his or her investigations.

Score Point 2: Apprentice

The student demonstrates an understanding of solving exponential and logarithmic equations. He or she uses mathematical reasoning at times during the task. He or she uses appropriate mathematical language some of the time. Student attempts to apply theoretical knowledge to the task but may be able to draw conclusions from his or her investigation.

Score Point 1: Novice

The student demonstrates a basic understanding of solving exponential and logarithmic equations. He or she uses appropriate mathematical language some of the time. Theoretical knowledge is extremely weak and many responses are irrelevant or illogical. He or she may fail to follow directions and has great difficulty in communicating his or her responses.

Score Point 0: Unsatisfactory

Student fails to make an attempt to complete the task and his or her responses are just an attempt to fill the page or restate the problem.

ANSWERS

Practice & Apply — Chapter 8

Lesson 8.1

1. $x = 4; y = 8$ **2.** $x = 5; y = 5$

3. $x = 6; y = 6\sqrt{3}$ **4.** $16\sqrt{3}$ **5.** 8   **6.** $2\sqrt{3}$

7. $\frac{100\sqrt{3}}{3}$ feet   **8.** 96 in.   **9.** 27 in.

Lesson 8.2

1. $x = 0.91; y = 0.42$ **2.** $x = -0.17; y = 0.98$

3. $x = -0.77; y = 0.64$ **4.** C   **5.** A   **6.** B

7. $\left(-\frac{\sqrt{3}}{2}, \frac{1}{2}\right)$ **8.** $\left(-\frac{\sqrt{3}}{2}, -\frac{1}{2}\right)$ **9.** $\left(\frac{\sqrt{3}}{2}, -\frac{1}{2}\right)$

For exercises 10–12, student answers may vary. Sample answers are given.

10. $-324°$ or $396°$ **11.** $300°$ or $-60°$

12. $-60°$ **13.** $65°$ west of north

14. $80°$ west of south   **15.** $20°$ east of south

16. $40°$ east of north

Lesson 8.3

1. $45°, 225°$ **2.** $210°; 330°$ **3.** $60°; 300°$

4. $0°; 180°$ **5.** $134°; 226°$ **6.** $50°; 230°$

7. $49°; 131°$ **8.** $102°; 258°$ **9.** $51°$

10. $76°$ **11.** $16°$ **12.** $9.6°$

13. $67°$ north of east   **14.** $35.5°$ **15.** $9°$

Lesson 8.4

1. The graphs have the same periods; there is no vertical shift or phase shift; the amplitude of the first graph is greater than the amplitude of the second.

2. $1, 2\pi, \frac{\pi}{2}$ **3.** $2, \pi, 0$

4.

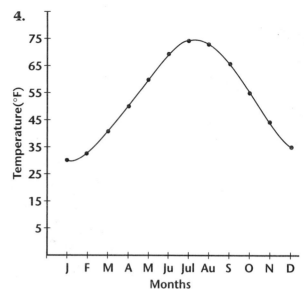

5. 43.8   **6.** 1 year   **7.** $y = 2 \sin \left(x - \frac{\pi}{4}\right)$

8. $y = \cos 3x$

Lesson 8.5

1. -0.1782 **2.** -0.1425 **3.** $\frac{\sqrt{2}}{2}$ **4.** -1

5.

Degrees	Radians	Sine	Tangent
60	1.0472	0.8660	1.7312
30	0.5236	0.5000	0.5774
15	0.2618	0.2588	0.2679
7.5	0.1309	0.1305	0.1316
3.75	0.0654	0.0654	0.0655
1.875	0.0327	0.0327	0.0327

6. $1, 4, 2\pi, -\pi$ **7.** $-2, 1, \frac{2\pi}{3}, \frac{\pi}{2}$

8. $f(x) = 4\sin x + 2$ **9.** $f(x) = 2\sin\left(x - \frac{\pi}{6}\right) + 1$

Lesson 8.6

1. $\frac{14\pi}{3}$ cm   **2.** 3π m^2 **3.** 4 m

4. length = 240.9 ft; area = 7225.7 ft^2

5. 687.2 in./min

6. larger: 11310 in./min; smaller 11310 in./min

7. 1.7 radians

ANSWERS

Lesson 8.7

1. domain: Dec. 22 to Dec. 22; range: $-23.5°$ to $23.5°$; period: one year

2. $23.5°$ 3. at the summer solstice

4. The sun is lower in the sky.

5. Check students' graphs.   6. 151,250 ft

7. π 8. Check students' graphs.

9. amplitude: 40; period: 12; vertical shift: 74

Enrichment — Chapter 8

Lesson 8.1

1. $\sin A = \cos A \dfrac{\sqrt{2}}{2}$; $\tan A = 1$; $45°$

2. $\sin A = \cos A = \dfrac{\sqrt{2}}{2}$; $\tan A = 1$; $45°$

3. $\sin A = \dfrac{1}{2}$; $\cos A = \dfrac{\sqrt{3}}{2}$; $\tan A = \dfrac{\sqrt{3}}{3}$; $30°$

4. $\sin A = \dfrac{\sqrt{3}}{2}$; $\cos A = \dfrac{1}{2}$; $\tan A = \sqrt{3}$; $60°$

5. $\sin A = \dfrac{\sqrt{3}}{2}$; $\cos A = \dfrac{1}{2}$, $\tan A = \sqrt{3}$; $60°$

6. $\sin A = \dfrac{1}{2}$; $\cos A = \dfrac{\sqrt{3}}{2}$; $\tan A = \dfrac{\sqrt{3}}{3}$; $30°$

7. $\sin A = \dfrac{\sqrt{3}}{2}$; $\cos A = \dfrac{1}{2}$; $\tan A = \sqrt{3}$; $60°$

8. $\sin A = \dfrac{1}{2}$; $\cos A \dfrac{\sqrt{3}}{2}$; $\tan A = \dfrac{\sqrt{3}}{3}$; $30°$

9. $\sin A = \cos A = \dfrac{\sqrt{2}}{2}$; $\tan A = 1$; $45°$

10. $\sin A = \dfrac{1}{2}$; $\cos A = \dfrac{\sqrt{3}}{2}$; $\tan A = \dfrac{\sqrt{3}}{3}$; $30°$

Lesson 8.2

1. 0.714   2. 0.908   3. 0.493   4. 0.903

5. 0.807   6. 0.978   7. 0.280   8. 0.998

9. 0.475   10. 0.760   11. 0.866   12. 1

13. no   14. yes   15. yes   16. no   17. yes

18. no   19. no   20. yes   21. yes   22. yes

23. no   24. no

Lesson 8.3

1. 0.7647   2. 0.8660   3. -1.7645

4. -0.6374 5. 0.9936   6. 0.4819

7. -4.6791 8. 0.6697   9. 0.4906

10. 0.2805   11. 0.9998   12. 0.7111

13. 0.2227   14. 0.9887   15. 0.3816

16. 1.0000   17. 0.8228   18. 0.5151

Lesson 8.4

1. $y = \sin x$; $y = \cos (x - 90)$

2. $y = \cos x$; $y = -\sin (x - 90)$

3. $y = \sin (x + 30)$; $y = \cos (x - 60)$

4. $y = \dfrac{1}{2} \sin x + 2$; $y = \dfrac{1}{2} \cos(x - 90) + 2$

5. $y = -\sin 3x$; $y = \cos (3x - 270)$

6. $y = \sin \left(\dfrac{1}{3}x\right) - 1$; $y = \cos \left(\dfrac{1}{3}x - 90\right) - 1$

Lesson 8.5

1. 250π rad/min   2. $66\dfrac{2}{3}\pi$ rad/min

3. 1000π rad/min   4. $\dfrac{8}{3}\pi$ rad/s

5. 156π rad/min   6. 125π rad/min

7. 720π rad/min   8. 25π rad/s

9. 10π rad/min   10. 9π rad/s

Lesson 8.6

1. 1.15 mi 2. 230.15 mi 3. 55.24 mi

4. 662.83 mi 5. 115.08 mi 6. 57.54 mi

7. 44.71 mi 8. 7.25 mi 9. 107.83 mi

10. 5.52 mi 11. 6.95 nautical mi

12. 1.74 nautical mi 13. 30.41 nautical mi

14. 65.17 nautical mi 15. 173.80 nautical mi

16. 2172.49 nautical mi 17. 1.56 nautical mi

18. 3.04 nautical mi 19. 16.46 nautical mi

20. 1464.78 nautical mi

Lesson 8.7

1. 49.1 ft 2. 38.4 m 3. 3.75 s

4. 0.4 s and 5.4 s 5. 0.1 s 6. 0.7 ft

Technology — Chapter 8

Lesson 8.1

1. The entries in columns D and E all equal 1.41421356. The entries in column F equal 1.

2. The entries in column D equal 0.4472136. The entries in column E equal 0.89442719. The entries in column F equal 2.

3. The entries in column D equal 0.31622777. The entries in column E equal 0.9486833. The entries in column F equal 3.

4. The entries in column D equal 0.24253563. The entries in column E equal 0.9701425. The entries in column F equal 4.

5. The entries in column D equal 0.19611614. The entries in column E equal 0.98058068. The entries in column F equal 5.

6. The entries in column D equal 0.16439899. The entries in column E equal 0.98639392. The entries in column F equal 6.

7. The entries in column D equal 0.14142136. The entries in column E equal 0.98994949. The entries in column F equal 7.

8. The entries in column D are equal. The entries in column E are equal. If $PQ = rOQ$, where $r > 0$, then the entries in column F equal r.

Lesson 8.2

1.

	A	B	C
1	23	0	23
2	113	0	113
3	203	0	203
4	293	0	293
5	383	1	23
6	473	1	113
7	563	1	203
8	653	1	293
9	743	2	23
10	833	2	113
11	923	2	203
12	1013	2	293
13	1103	3	23
14	1193	3	113
15	1283	3	203
16	1373	3	293
17	1463	4	23
18	1553	4	113
19	1643	4	203
20	1733	4	293

2. For a given angle measure in column A, the corresponding entry in column B tells how many complete revolutions the terminal side of the angle makes.

3. 23°, 383°, 743°, 1103°, and 1463°

4. 23° + (360°)n, where n is a positive integer.

5. Find the largest integer multiple of 360° that is less than *s*. Then subtract this amount from *s*. The result is *r*.

Lesson 8.3

1. $a = 9.64$; $b = 11.49$

2. $a = 13.24$; $b = 7.04$

3. $a = 5.56$; $b = 22.32$

4. $a = 1.28$; $b = 0.23$

5. $a = 1.74$; $b = 9.85$

6. $a = 150.14$; $b = 45.90$

7. $a = 2.27$; $b = 4.46$

8. $a = 7.80$; $b = 0.14$

9. $a = 186.15$; $b = 156.20$

Lesson 8.4

1.

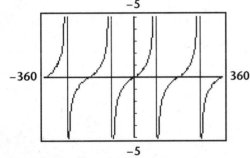

2. The display shows congruent but disjoint branches.

3. The tangent equals 0 when *x* is an even multiple of 90°; $x = n(90°)$, where *n* is an even integer.

4. The tangent is undefined when *x* is an odd multiple of 90°; $x = n(90°)$, where *n* is an odd integer.

5. The domain is all real numbers except the odd multiples of 90°. The range is the set of all real numbers.

6. The period is 180°.

7. No; only functions with a maximum and minimum can have an amplitude.

8. tan 1170° is not defined, since 1170° is an odd multiple of 90°.

Lesson 8.5

1. 1.047 radians   **2.** 24.06°

3. 0.175 radians   **4.** 91.67°

5. 1.257 radians   **6.** 114.59°

7. 0.5235 radians   **8.** ≈ 46°

9. 0.78525 radians   **10.** ≈ 63°

11. $360° < \theta < 720°$

Lesson 8.6

1.

	A	B	C
1	RADIUS	REV/MIN	LIN SPD
2	0.0	15.5	0
3	0.5	15.5	48.694645
4	1.0	15.5	97.38929
5	1.5	15.5	146.083935
6	2.0	15.5	194.77858
7	2.5	15.5	243.473225
8	3.0	15.5	292.16787
9	3.5	15.5	340.862515
10	4.0	15.5	389.55716
11	4.5	15.5	438.251805
12	5.0	15.5	486.94645

ANSWERS

2.

	A	B	C
1	RADIUS	REV/MIN	LIN SPD
2	0.0	24.2	0
3	0.5	24.2	76.026478
4	1.0	24.2	152.052956
5	1.5	24.2	228.079434
6	2.0	24.2	304.105912
7	2.5	24.2	380.13239
8	3.0	24.2	456.158868
9	3.5	24.2	532.185346
10	4.0	24.2	608.211824
11	4.5	24.2	684.238302
12	5.0	24.2	760.26478

3.

	A	B	C
1	RADIUS	REV/MIN	LIN SPD
2	22.7	0.0	0
3	22.7	0.5	71.314093
4	22.7	1.0	142.628186
5	22.7	1.5	213.942279
6	22.7	2.0	285.256372
7	22.7	2.5	356.570465
8	22.7	3.0	427.884558
9	22.7	3.5	499.198651
10	22.7	4.0	570.512744
11	22.7	4.5	641.826837
12	22.7	5.0	713.14093
13	22.7	5.5	784.455023
14	22.7	6.0	855.769116

4.

	A	B	C
1	RADIUS	REV/MIN	LIN SPD
2	130.2	0.0	0
3	130.2	0.5	409.035018
4	130.2	1.0	818.070036
5	130.2	1.5	1227.10505
6	130.2	2.0	1636.14007
7	130.2	2.5	2045.17509
8	130.2	3.0	2454.21011
9	130.2	3.5	2863.24513
10	130.2	4.0	3272.28014
11	130.2	4.5	3681.31516
12	130.2	5.0	4090.35018
13	130.2	5.5	4499.3852
14	130.2	6.0	4908.42022

5. The linear speed increases.

6. The linear speed increases.

Lesson 8.7

1. $y = -7 \cos\left(\frac{\pi t}{4}\right)$ **2.** $y = 7 \cos\left(\frac{\pi t}{4}\right)$

3. $y = -4.5 \cos\left(\frac{\pi t}{3}\right)$ **4.** $y = -7 \cos\left(\frac{\pi t}{3}\right)$

5. $y = -a \cos\left(\frac{2\pi t}{c}\right)$

Lesson Activities — Chapter 8

Lesson 8.1

1. Each triangle has an area of 2 square units.

2. The four triangles have the same area.

3.

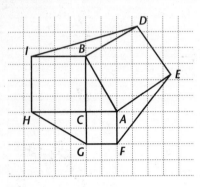

4. Each triangle has an area of $2\sqrt{3}$ square units.

5. The four triangles have the same area.

6.

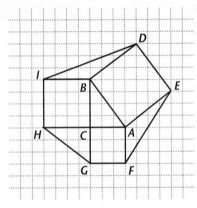

7. Each triangle has an area of 6 square units.

8. The areas of the triangles are equal.

9. The areas of the triangles formed by joining the vertices of the squares are equal to each other and the area of the original right triangle.

Lesson 8.2

1. Answers will vary. Check student angles.

2. Answers will vary. **3–4.** Answers will vary.

5. The sum of the ratios is 1.

6–7. Answers will vary.

Lesson 8.3

	$S(x)$	$\sin d$
1.	0.1045284633	0.1045284633
2.	0.7193398003	0.7193398003
3.	0.4383118415	0.4383711468
4.	−0.4723433697	−0.4694715628

	$C(x)$	$\cos d$
5.	0.9945218954	0.9945218954
6.	0.6946583703	0.6946583705
7.	−0.8990793959	−0.8987940463
8.	−0.893135552	−0.8829475929

9. The tan of any angle can be approximated by using $\frac{S(x)}{C(x)}$.

Lesson 8.4

1.

Xmax	360	34200	68040	101880
Cycles	1	1	1	1

Xmax	135720	169560
Cycles	1	1

2.

Xmax	33840	34200	34560	34920
Cycles	0	1	2	3

Xmax	35280	35640
Cycles	4	5

3. $360 + 94(360)n$, where n is a positive integer less than 10.

4. $(94 + n)360$, where n is the number of cycles.

5–6. Answers will vary. Check students' responses.

ANSWERS

Lesson 8.5

1–4.

Point	First	Second	Third	Fourth
y	0.54	0.86	0.65	0.79

Point	Fifth	Sixth	Seventh
y	0.70	0.76	0.72

5. $\cos(1) = 0.54$; $\cos(\cos(1)) = 0.86$;
$\cos(\cos(\cos(1))) = 0.65$;
$\cos(\cos(\cos(\cos(1)))) = 0.79$;
$\cos(\cos(\cos(\cos(\cos(1))))) = 0.709$;
$\cos(\cos(\cos(\cos(\cos(\cos(1)))))) = 0.76$;
$\cos(\cos(\cos(\cos(\cos(\cos(\cos(1))))))) = 0.72$;
$\cos(\cos(\cos(...(1)))) = 0.74$

Lesson 8.6

1. ≈ 896 miles; ≈ 1100 miles

2. 881 miles; 1073 miles; 1233 miles

3. As the distance above the Earth increases, the arc length increases.

4. 4,976,482 mi²; 2,488,241 mi²; 3,732,362 mi²

Lesson 8.7

1. Check students' graphs.

2. At 400 feet, the ball is about 16 feet high, so it does not clear the fence.

3. The ball is about 21 feet high and about 381 feet from the batter. An outfielder could not catch the ball.

4.

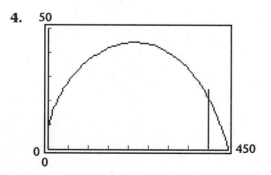

5. Yes, the ball hits the fence.

6. No, the ball does not hit the fence.

Assessment — Chapter 8

Assessing Prior Knowledge 8.1

1. $\sqrt{2}$ **2.** $\sqrt{3}$

Quiz 8.1

1. $x = 5\sqrt{3}$ cm
$y = 10$ cm

2. $x = 6\sqrt{2}$ cm
$y = 6\sqrt{2}$ cm

3. $x = 3\sqrt{2}$ in.
$y = 6$ in.

4. $x = \frac{8\sqrt{3}}{3}$ in.
$y = \frac{16\sqrt{3}}{3}$ in.

5. $\frac{5\sqrt{6}}{3}$ cm   **6.** $100\sqrt{3}$ m²

Assessing Prior Knowledge 8.2

1. $\sqrt{29}$ **2.** 1   **3.** 1

Quiz 8.2

1. 120°

2.

3.

HRW material copyrighted under notice appearing earlier in this work.

ANSWERS

4.

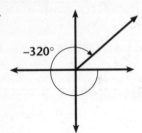

5.

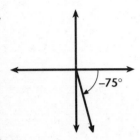

6.

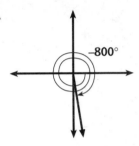

7.

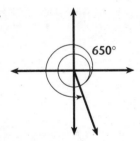

8. $x = -0.71$
$y = -0.71$

9. $x = -0.50$
$y = 0.87$

10. $x = 0.77$
$y = -0.64$

11. $x = 0.42$
$y = 0.91$

Assessing Prior Knowledge 8.3

1. $f^{-1}(x) = \frac{x + 8}{3}$ **2.** $f^{-1}(x) = \ln\left(\frac{x}{2}\right)$

Quiz 8.3

1. $30°, 210°$ **2.** $60°, 300°$

3. $60°, 120°$ **4.** $120°, 300°$

5. Find $\cos^{-1}0.92$ or $\sin^{-1}0.40$ or $\tan^{-1}\left(\frac{0.40}{0.92}\right)$

6. Yes. When $\theta = 330°$, the inverse relation is true.

7. No. Values of $\sin^{-1}0$ and $\cos^{-1}0$ are always $0°$ and $90°$ respectively.

8. 1

Assessing Prior Knowledge 8.4

1. $(-1, -2)$ **2.** $(3, 4)$

Quiz 8.4

1. $0, 1, 90°, 0°$

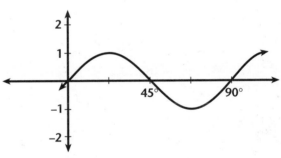

2. $3, 1, 360°, 0°$

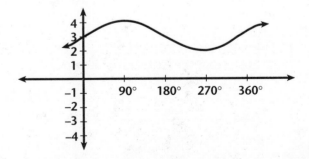

ANSWERS

3. 0, 5, 180°, 0°

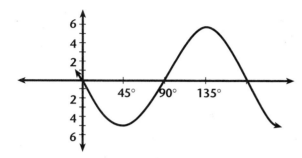

4. −2, 3, 720°, −90°

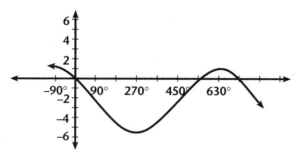

5. $f(\theta) = 4\sin 2\theta$

6. $f(\theta) = -1 - 3\cos\frac{1}{2}\theta$ or $-1 + 3\sin\frac{1}{2}(\theta - 180)$

Mid-Chapter Assessment

1. c   **2.** b   **3.** c   **4.** d   **5.** 10 in.

6. 249 in.²

7.

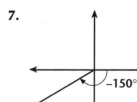

8.

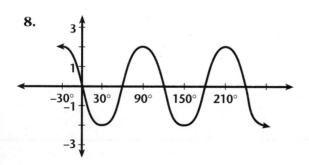

Assessing Prior Knowledge 8.5

1. 57.30   **2.** 0.017

Quiz 8.5

1. $\frac{\sqrt{3}}{2}$ **2.** undefined   **3.** $-\frac{\sqrt{2}}{2}$ **4.** 0

5. $\frac{7\pi}{18} = 1.22$ **6.** $\frac{8\pi}{9} = -2.79$

7. −270°

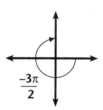

8. 85.9°

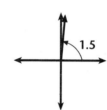

9. 0, 3, 4, 0   **10.** $-2, 1, 6\pi, -\frac{\pi}{3}$ **11.** 0.48

12. −1

Assessing Prior Knowledge 8.6

1. 2π **2.** 3π **3.** 4.6π

Quiz 8.6

1. 25.6 m   **2.** 440 m²   **3.** 166 m²

4. 36.4 m   **5.** 40.14 ft   **6.** 9.56 ft/sec

Assessing Prior Knowledge 8.7

up 5 units, $-\frac{\pi}{3}\left(\frac{\pi}{3}\right.$ units to the left), vertical

stretch of 2 (3 to 7), $\frac{2\pi}{3}$

ANSWERS

Quiz 8.7

1. max: $y = 8$ at $x = 0, 4, \ldots$; min $y = -2$ at $x = -2, 2, \ldots$

2. 3   **3.** 3   **4.** 0, 4   **5.** 4 radians   **6.** 5

7. 3

Chapter Assessment, Form A

1. b   **2.** c   **3.** c   **4.** c   **5.** b   **6.** a   **7.** b

8. d   **9.** b   **10.** a   **11.** d   **12.** c

Chapter Assessment, Form B

1. $4\sqrt{3}$ cm   **2.** $\frac{225\sqrt{3}}{4}$ in.²

3.

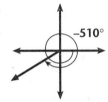

4. -0.94 **5.** $217°$ **6.** $150°, 210°$ **7.** $240°$

8. $-1, 2, \frac{3\pi}{2}, \frac{\pi}{2}$ **9.** $f(\theta) = 1 - 2\sin\frac{1}{2}\theta$

10. -1

11.

12. 0.31 **13.** 7.2 m   **14.** $x \approx 2.95; x \approx 5.05$

Alternative Assessment — Chapter 8

Form A

1. Period: $360°$; Amplitude: 1; $45°$ to the right

2. Period: $360°$; Amplitude: 2; $60°$ to the left

3. Period: $360°$; Amplitude: 1; Vertical shift up of 2

4. The graph of $f(x) = \cos(x - d)$ and $\sin(x - d)$ results by shifting the graph of $f(x) = \cos x$ and $f(x) = \sin x$ to the right d units if $d > 0$ or to the left $|d|$ units if $d < 0$.

5. Domain: all real numbers; range: $-1 \le y \le 1$; period $180°$

6. Domain: all real numbers; range: $-1 \le y \le 1$; period $720°$

7. Domain: all real numbers; range: $-5 \le y \le 5$; period $1080°$

8. Answers may vary. $f(x) = \sin 3x$

Score Point 4: Distinguished

The student demonstrates a comprehensive understanding of graphs of trigonometric functions. The student uses perceptive, creative, and complex mathematical reasoning throughout the task. He or she is able to use sophisticated, precise, and appropriate mathematical language throughout the task. Theoretical knowledge is apparent and applied to concrete situations as the student successfully demonstrates a comprehensive understanding of core concepts throughout the task.

Score Point 3: Proficient

The student demonstrates a broad understanding of graphs of trigonometric functions. The student uses perceptive mathematical reasoning most of the time. He or she is able to use precise and appropriate mathematical language most of the time. Theoretical knowledge is apparent and applied to concrete situations as the student attempts to draw conclusions based on his or her investigations.

Score Point 2: Apprentice

The student demonstrates an understanding of graphs of trigonometric functions. He or she uses mathematical reasoning at times during the task. He or she uses appropriate mathematical language some of the time. Student attempts to apply theoretical knowledge to the task but may be able to draw conclusions from his or her investigation.

Score Point 1: Novice

The student demonstrates a basic understanding of graphs of trigonometric functions. He or she uses appropriate mathematical language some of the time. Theoretical knowledge is extremely weak and many responses are irrelevant or illogical. He or she may fail to follow directions and has great difficulty in communicating his or her responses.

Score Point 0: Unsatisfactory

Student fails to make an attempt to complete the task and his or her responses are just an attempt to fill the page or restate the problem.

Form B

1. The period is 1.2 seconds. The period represents the time between contractions of the heart.

2. The diastolic pressure is 100 mm.

3. The diastolic pressure is 110 mm at mercury readings of 0.4, 0.8, 1.6, 2.0, 2.8, 3.2, and so on.

4. The amplitude is 40 mm which represents the difference between the maximum BP of 120 mm and the minimum or resting BP of 80 mm.

5. The vertical shift is 100 mm.

Score Point 4: Distinguished

The student demonstrates a comprehensive understanding of applications of trigonometric functions. The student uses perceptive, creative, and complex mathematical reasoning throughout the task. He or she is able to use sophisticated precise, and appropriate mathematical language throughout the task. Theoretical knowledge is apparent and applied to concrete situations as the student successfully demonstrates a comprehensive understanding of core concepts throughout the task.

Score Point 3: Proficient

The student demonstrates a broad understanding of applications of trigonometric functions. The student uses perceptive mathematical reasoning most of the time. He or she is able to use precise and appropriate mathematical language most of the time. Theoretical knowledge is apparent and applied to concrete situations as the student attempts to draw conclusions based on his or her investigations.

Score Point 2: Apprentice

The student demonstrates an understanding of applications of trigonometric functions. He or she uses mathematical reasoning at times during the task. He or she uses appropriate mathematical language some of the time. Student attempts to apply theoretical knowledge to the task but may be able to draw conclusions from his or her investigation.

Score Point 1: Novice

The student demonstrates a basic understanding of applications of trigonometric functions. He or she uses appropriate mathematical language some of the time. Theoretical knowledge is extremely weak and many responses are irrelevant or illogical. He or she may fail to follow directions and has great difficulty in communicating his or her responses.

Score Point 0: Unsatisfactory

Student fails to make an attempt to complete the task and his or her responses are just an attempt to fill the page or restate the problem.

Practice & Apply — Chapter 9

Lesson 9.1

1. 4   2. $y = \dfrac{4}{x^3}$

3.
x	2	4	10
y	0.5	0.0625	0.004
;

Check students' graphs.

4.
x	3	6	8	a	$12a$
y	4	2	1.5	$\dfrac{12}{a}$	$\dfrac{1}{a}$

ANSWERS

5. $k = 8; y = \dfrac{8}{x}$ **6.** $k = 8; y = \dfrac{8}{x^2}$

7. $k = 64; y = \dfrac{64}{x^3}$

8. $y = \dfrac{4.67}{x^2}$

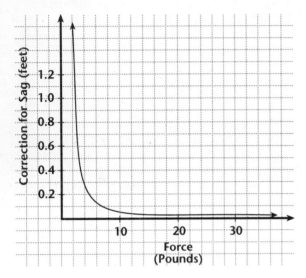

9. 0.0238   **10.** sag correction decreases

Lesson 9.2

1. D   **2.** C   **3.** A   **4.** B

For exercises 5–8, check students' graphs.

5. Domain: all real numbers except 1; Range: all real numbers except 0; Asymptotes: $x = 1, y = 0$

6. Domain: all real numbers except 0; Range: all real numbers except 1; Asymptotes: $x = 0, y = 1$.

7. Domain: all real numbers except 0; Range: all real numbers except 1; Asymptotes: $x = 0, y = 1$.

8. Domain: all real numbers except -1; Range: all real numbers except 1; Asymptotes: $x = -1, y = 1$.

9. $C = \dfrac{0.5}{5 + x}$

10.

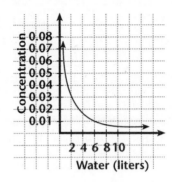

11. 5%

Lesson 9.3

1. even   **2.** odd   **3.** neither

4.

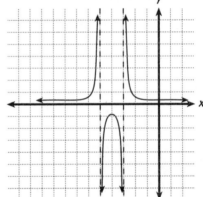

5.

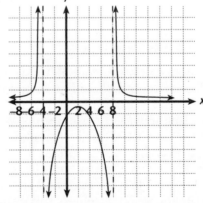

ANSWERS

6.

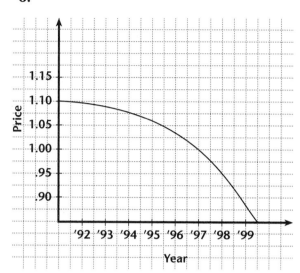

7. All years beginning with 1991

8. The price in 1991 was approximately $1.10; the price in 1994 was approximately $1.07.

9. $0.91

Lesson 9.4

1. $\dfrac{x+4}{x-3}$ **2.** $\dfrac{2x+1}{2x+3}$ **3.** $\dfrac{-2}{x-1}$ **4.** $\dfrac{x+5}{x^2}$

5. asymptotes: $x = -6$, $x = -4$, $y = 1$; domain: all real numbers except -6 and -4

6. asymptotes: $x = 2$, $x = -1$, $y = 1$; domain: all real numbers except 2 and -1

7. asymptotes: $x = 1$, $x = -4$, $y = 0$; domain: all real numbers except 0, 1, and -4.

8. asymptotes: $x = 0$, $y = 1$; domain all real numbers except 0 and 2.

9. $0.92 **10.** $0.58 **11.** $0.42

12. $0.26 **13.** $0.18

14. The average purchasing power of the dollar decreases more slowly as the number of years since 1970 increases, but will never reach 0.

Lesson 9.5

1. $\dfrac{x^2 + 7x - 10}{x^2 - 25}$ **2.** $\dfrac{x+6}{x^2 - 25}$ **3.** $\dfrac{-x^2 + x + 5}{x^2 + 5x}$

4. $\dfrac{x^2 + 9x + 25}{x^2 + 5x}$ **5.** $x = 1.8$ or $x = -2.3$

6. $x = 4$ or $x = 1$ **7.** $x = 2.3$ or $x = -0.34$

8. $x = 13$

9. smaller = 4.58 m; larger = 7.42 m

10. $x = 4$ **11.** $x = 1$ or $x = -3$

12. about 24%

Enrichment — Chapter 9

Lesson 9.1

1. $y = 48$ **2.** $x = \pm 4\sqrt{2}$ **3.** $y = 64$

4. $y = \dfrac{3}{32}$ **5.** $y = \dfrac{32}{3}$ **6.** $y = 12$ **7.** $y = \dfrac{4}{3}$

8. $y = 128$

Lesson 9.2

1. vertical asymptote: $x = 4$; undefined at $x = -3$

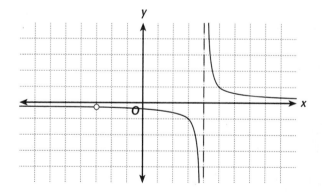

---

2. vertical asymptote: $x = -5$; undefined at $x = 4$

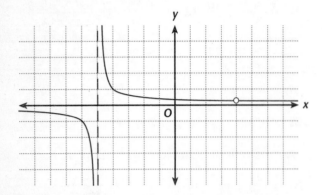

3. vertical asymptote: $x = 1$; undefined at $x = 6$

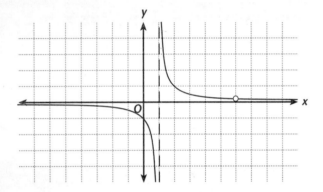

4. vertical asymptote: $x = -1$; undefined at $x = -2$

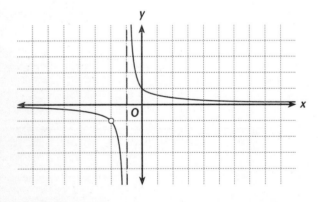

Lesson 9.3

1. 13 h   **2.** 11.82 h   **3.** 10.83 h

4. 16.25 h   **5.** 10 h   **6.** 14.44 h

7. 0.01 m/s^2 **8.** 1 m/s^2 **9.** 0.02 m/s^2

10. 0.25 m/s^2 **11.** 100 m/s^2 **12.** 1000 m/s^2

13. 39.48   **14.** 157.91   **15.** 9.87

16. 631.65   **17.** 394,784.18   **18.** 0.39

19. 13.33   **20.** 13.64   **21.** 23.08

22. 10.91   **23.** 21.05   **24.** 13.04

Lesson 9.4

1. horizontal; $y = 2$ **2.** oblique; $y = x + 5$

3. horizontal; $y = 0$ **4.** horizontal; $y = \frac{3}{2}$

5. oblique; $y = 3x - 14$ **6.** horizontal; $y = 0$

7. horizontal; $y = 1$ **8.** horizontal; $y = 3$

9. horizontal; $y = 0$ **10.** horizontal; $y = 0$

11. oblique; $y = x - 2$ **12.** oblique; $y = 5x + 18$

Lesson 9.5

1. $3 < x < \frac{13}{3}$ **2.** $-10 < x < -6$

3. $x > -2$ **4.** $x > -1$

5. $x < -1$ or $-\frac{1}{3} < x < 1$

6. $-3 < x < 4$ or $x > \frac{15}{2}$

7. $x < -1$ or $x > 1$

8. $-2 < x < -1$ or $3 < x < 9$

Technology — Chapter 9

Lesson 9.1

1. $\dfrac{\dfrac{1}{x+1} - \dfrac{1}{x}}{(x+1) - x} = \dfrac{1}{x+1} - \dfrac{1}{x}$

2. Cell A1 contains 1. Cell A2 contains 1 + A1. Cell B1 contains 1/(A1+1)−1/A1. Then use FILL DOWN to fill 11 rows.

3. The slopes of $\overline{MN}$ form a sequence of negative numbers that approach 0, the slope of a horizontal line.

ANSWERS

4.

	A	B
1	1	−1.0000000
2	2	−0.3333333
3	3	−0.1666667
4	4	−0.1000000
5	5	−0.0666667
6	6	−0.0476190
7	7	−0.0357143
8	8	−0.0277778
9	9	−0.0222222
10	10	−0.0181818

5.

	A	B
1	1	−1.5000000
2	2	−0.5000000
3	3	−0.2500000
4	4	−0.1500000
5	5	−0.1000000
6	6	−0.0714286
7	7	−0.0535714
8	8	−0.0416667
9	9	−0.0333333
10	10	−0.0272727

6.

	A	B
1	1	−2.0000000
2	2	−0.6666667
3	3	−0.3333333
4	4	−0.2000000
5	5	−0.1333333
6	6	−0.0952381
7	7	−0.0714286
8	8	−0.0555555
9	9	−0.0444444
10	10	−0.0363636

7.

	A	B
1	1	3
2	2	5
3	3	7
4	4	9
5	5	11
6	6	13
7	7	15
8	8	17
9	9	19
10	10	21

The entries in column B form a sequence of positive numbers that increase without bound. The graph gets straighter vertically, since a vertical line has no slope.

Lesson 9.2

1.

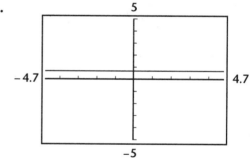

2.

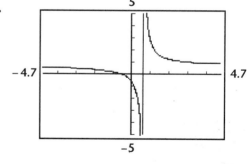

ANSWERS

3.

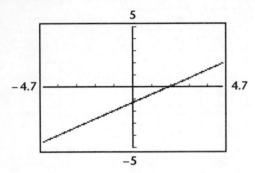

4.

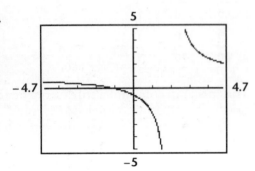

5.

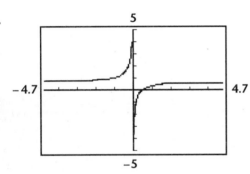

6.

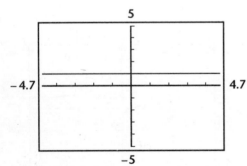

7.

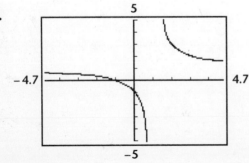

8.

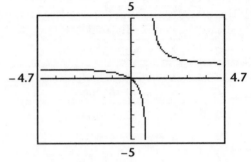

9. The graph is the horizontal line with equation $y = \frac{b}{d}$.

10. The graph is a slanted line with slope $\frac{a}{d}$ and y-intercept $\frac{b}{d}$.

11. The graph is the horizontal line with equation $y = \frac{a}{c}$ but with a hole in it at $(0, \frac{a}{c})$.

12. The graph is a hyperbola with horizontal asymptote $y = \frac{a}{c}$ and vertical asymptote $x = -\frac{d}{c}$.

Lesson 9.3

1-7.

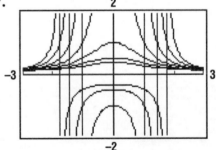

8. If $c > 0$, the graph is a hill symmetric about the y-axis. Its peak is $(0, \frac{1}{c})$ and the hill trails off to the y-axis as x increases without bound.

9. If $c = 0$, the graph is a pair of branches in the first and second quadrants symmetric about the y-axis. Each branch approaches the y-axis as x approaches 0 and the x-axis if x increases without bound.

ANSWERS

10. If $c < 0$, the graph is a pair of branches in the first and second quadrants symmetric about the y-axis. Each branch approaches $\sqrt{c}$ and $-\sqrt{c}$ as x approaches $\sqrt{c}$ and $-\sqrt{c}$ from the right and from the left, respectively, and the x-axis as x increases without bound. In addition, there is a U shaped hill between the lines $x = \sqrt{c}$ and $x = -\sqrt{c}$ whose peak has coordinates $(0, \frac{1}{c})$. As c gets smaller, the hill gets flatter at the top.

Lesson 9.4

1.

	A	B
1	−4.00	−4.2500000
2	−3.75	−4.0166667
3	−3.50	−3.7857143
4	−3.25	−3.5576923
5	−3.00	−3.3333333
6	−2.75	−3.1136364
7	−2.50	−2.9000000
8	−2.25	−2.6944444
9	−2.00	−2.5000000
10	−1.75	−2.3214286
11	−1.50	−2.1666667
12	−1.25	−2.0500000
13	−1.00	−2.0000000
14	−0.75	−2.0833333
15	−0.50	−2.5000000
16	−0.25	−4.2500000
17	0.00	#DIV/0!
18	0.25	4.25000000
19	0.50	2.50000000
20	0.75	2.08333333
21	1.00	2.00000000
22	1.25	2.05000000
23	1.50	2.16666667
24	1.75	2.32142857
25	2.00	2.50000000
26	2.25	2.69444444
27	2.50	2.90000000
28	2.75	3.11363636
29	3.00	3.33333333
30	3.25	3.55769231
31	3.50	3.78571429
32	3.75	4.01666667
33	4.00	4.25000000

The table indicates that the slope is negative and increasing until $x = -1$ at which point the slope begins to decrease without bound. At $x = 0$, the slope is undefined. Between $x = 0$ and $x = 1$, the slope is positive but decreasing. At $x = 1$, it reaches its positive minimum. For $x > 1$, the slope begins to increase without bound.

2.

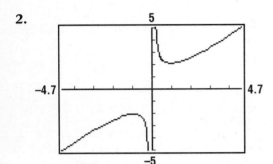

Yes. The graph confirms that the slope is negative and increasing until $x = -1$ at which point the slope decreases without bound. At $x = 0$, the slope is undefined. Between $x = 0$ and $x = 1$, the slope is positive but decreasing. At $x = 1$, it reaches it positive minimum. For $x > 1$, the slope begins to increase without bound.

3.

	A	B
1	−4.00	−3.7500000
2	−3.75	−3.4833333
3	−3.50	−3.2142857
4	−3.25	−2.9423077
5	−3.00	−2.6666667
6	−2.75	−2.3863636
7	−2.50	−2.1000000
8	−2.25	−1.8055556
9	−2.00	−1.5000000
10	−1.75	−1.1785714
11	−1.50	−0.8333333
12	−1.25	−0.4500000
13	−1.00	0.00000000
14	−0.75	0.58333333
15	−0.50	1.50000000
16	−0.25	3.75000000
17	0.00	#DIV/0!
18	0.25	−3.7500000
19	0.50	−1.5000000
20	0.75	−0.5833333
21	1.00	0.00000000
22	1.25	0.45000000
23	1.50	0.83333333
24	1.75	1.17857143
25	2.00	1.50000000
26	2.25	1.80555556
27	2.50	2.10000000
28	2.75	2.38636364
29	3.00	2.66666667
30	3.25	2.94230769
31	3.50	3.21428571
32	3.75	3.48333333
33	4.00	3.75000000

The table indicates that the slope is negative and increasing until $x = -1$ at which point the slope becomes positive and increases without bound. At $x = 0$, the slope is undefined. Between $x = 0$ and $x = 1$, the slope is negative and increasing. At $x = 1$, it becomes positive and increases without bound.

ANSWERS

4.

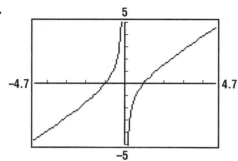

The graph confirms that the slope is negative and increasing until $x = -1$ at which point the slope is zero and thereafter becomes positive and increases without bound. At $x = 0$, the slope is undefined. Between $x = 0$ and $x = 1$, the slope is negative and increasing. At $x = 1$, it is zero and thereafter becomes positive and increases without bound.

5. The equation $f(x) = 0$ has roots $x = c$ and $x = d$ as long as c and d are nonzero.

Lesson 9.5

1. $\dfrac{18x - 4}{9x^2 - 9x + 2}$ **2.** $\dfrac{x + 5}{x + 1}$ **3.** $\dfrac{2x - 8}{6x^2 + x - 2}$

4. $\dfrac{-6x^2 - 2x - 1}{2x^2 + 5x - 3}$ **5.** $x = \dfrac{2}{5}$, or 0.4

6. $x = \dfrac{3}{5}$, or 0.6 **7.** $x = 5$, or -5

8. no solution

Lesson Activities — Chapter 9

Lesson 9.1

1. Check students' graphs. **2.** $y = 350x^{-1}$

3. The differences are all 0s, indicating that the curve exactly fits the data points.

4. The 70th punch mark.

5.

p	1	2	5	10	25
c	280	140	56	28	11.20

6. $y = 280x^{-1}$

7. The cost of each trip on the discounted pass is always cheaper.

Lesson 9.2

1. $\sqrt{10} \approx 3.1622777$

2. All the estimates converge to 3.1622777.

3. 1.4142136

4. $\sqrt{17} \approx 4.1231056$; $\sqrt{30} \approx 5.4772256$

5. The x-coordinate of the intersection point(s) represents an approximation of $\pm \sqrt{5}$ or ± 2.23.

Lesson 9.3

1. Check students' graphs.

2. Ymin $= -1$, Ymax $= 1$

3. Check students' graph. Ymin $= -0.2$, Ymax $= 0.2$

4. Possible answers: Xmin $= -4$, Xmax $= 2$, Xscl $= 0.6$, Ymin $= -1$, Ymax $= 1$, Yscl $= 1$

Lesson 9.4

1. horizontal asymptote: $y = 1$; $\left(\dfrac{7}{3}, 1\right)$

2. horizontal asymptote: $y = 2$; $(0, 2)$

3. The rational function does not cross its horizontal asymptote, $y = 1$.

4. The rational function does not cross its horizontal asymptote, $y = 1$.

5. No, the function does not always cross its horizontal asymptote.

6. horizontal asymptote: $y = 0$; $(-3, 0)$

7. horizontal asymptote: $y = 0$; $(2, 0)$ and $(1, 0)$

8. horizontal asymptote: $y = 0$; $(1, 0)$

9. horizontal asymptote: $y = 0$; $(1, 0)$

10. The function has a hole at $x = 2$. Since a hole in a graph is a removable discontinuity, the function simplifies to $\frac{1}{(x + 2)}$. The horizontal asymptote is $y = 0$. The function does not cross its horizontal asymptote.

11. A rational function crosses the horizontal asymptote when the degree of the numerator is one degree less than that of the denominator. As x approaches the point of intersection on the left and right sides, the limit of the function approaches the asymptote.

Lesson 9.5

1. The cobweb converges to the golden ratio 0.61903399.

2. All values converge to the golden ratio.

3. As the terms increase, the reciprocal of the golden ratio, 1.61803398875 is approached. 15th: $\frac{610}{377} \approx 1.61803713528$; 20th: $\frac{6765}{4181} \approx 1.61803396317$; 30th: $\frac{832040}{514229} \approx 1.61803398875$

Assessment — Chapter 9

Assessing Prior Knowledge 9.1

1. 18   2. $\frac{1}{20}$

Quiz 9.1

1. Answers may vary. Possible answer: $xy = 8$

2. 54   3. $x^3y = 54$ 4. $x = 3\sqrt[3]{10}$; $y = 0.25$

5. 100 lb   6. $5\frac{5}{8}$ h

Assessing Prior Knowledge 9.2

1. $-\frac{1}{4}, -\frac{1}{2}$, undefined, $\frac{1}{2}, \frac{1}{4}$

2. $\frac{2}{5}, -4$, undefined, $7, \frac{13}{5}$

Quiz 9.2

1. Domain: all real numbers except $x = -4$
Range: all real numbers except $y = 0$

Asymptotes: $y = 0$
$x = -4$

2. Domain: all real numbers except $x = 2$
Range: all real numbers except $y = 1$

Asymptotes: $y = 1$
$x = 2$

3. $16\frac{2}{3}\%$ 4. $\approx 23.1\%$ 5. $C(x) = \frac{6}{20 + x}$

6. 24% acid

Assessing Prior Knowledge 9.3

1. $(x + 1)(x - 2)$ 2. $(x + 3)(2x - 1)$

3. $(2x + 1)(x - 5)$

Quiz 9.3

1. Domain: all real numbers except $x = -2$ and $x = -3$

2. neither   3. neither

4. Answers may vary. Possible answer:
$f(x) = \frac{2x + 3}{x - 3}$

5. Check students' graphs. Possible answer: The graph has a vertical asymptote at $x = -5$, a horizontal asymptote at $y = 0$, and no maximum or minimum points. As x approaches -5 from the right or left, the value of the function increases.

6. Check students' graphs. Possible answer: The graph has a vertical asymptote at $x = 2$, a horizontal aymptote at $y = 0$, and no maximum or minimum points. The value of the function increases as x approaches 2 from either side.

ANSWERS

7. Check students' graphs. Possible answer: The graph has vertical asymptotes at $x = 3$ and $x = -3$; horizontal asymptote at $y = 0$; a maximum value of $-\frac{1}{9}$ in the interval $-3 < x < 3$; as x approaches -3 from the left, the function increases; the function also increases for $x < 0$, and decreases for $x > 0$.

8. Check students' graphs. Possible answer: The graph has a vertical asymptote of $x = -1$ and a horizontal aymptote of $y = 0$ and no maximum or minimum points. The function is always decreasing.

Mid-Chapter Assessment

1. c **2.** d **3.** a **4.** b **5.** d **6.** even

7. Answers will vary. Possible answer:
$$f(x) = \frac{x + 5}{x + 3}$$

8.

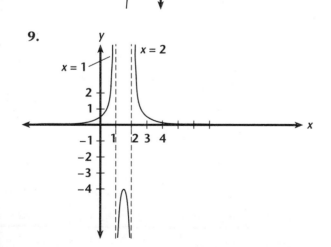

9.

Assessing Prior Knowledge 9.4

1. $2 + \dfrac{4x + 5}{x^2 - x - 2}$ **2.** $2 + \dfrac{4x - 1}{2x^2 + 3}$

Quiz 9.4

1. $x = -2$ **2.** $x = \pm 2$

3. $\dfrac{x + 5}{x - 2}$ for $x \neq 5$ and $x \neq 2$

4. $\dfrac{1}{x + 2}$ for $x \neq -2$

5. Vertical asymptote: $x = -1$
Horizontal asymptote: $y = 0$

Domain: all real numbers except $x \neq -1$

6. Vertical asymptote: $x = 2$, $x = -3$
Horizontal asymptote: $y = 1$

Domain: all real numbers except $x \neq 2$, $x = -3$, and $x = -4$

7. Yes; $x = -4$

Assessing Prior Knowledge 9.5

1. $\dfrac{31}{24}$ **2.** $\dfrac{2x^2 + 2x + 2}{x^2 + x}$

Quiz 9.5

1. $\dfrac{x^2 + 6x - 9}{(x + 3)(x - 3)}$ **2.** $\dfrac{2x + 9}{x^2 - 16}$

3. $x = 0.8, -1.3$ **4.** $x = \pm 3.8$

5. $x = 1.7, 1.5$ **6.** $x = 8, 0$ **7.** $x = 15\frac{1}{2}$

8. $x = 4.0, -3.5$ **9.** $2\frac{2}{9}$ h

Chapter Assessment, Form A

1. b **2.** c **3.** c **4.** a **5.** c **6.** d **7.** b

8. a **9.** c **10.** c **11.** b **12.** d **13.** c

14. a **15.** c **16.** b

ANSWERS

Chapter Assessment, Form B

1. Domain: all real numbers except $x \neq -4$
   Range: all real numbers except $y \neq 0$

2. Horizontal $y = 1$
   Vertical: $x = -3$, $x = -4$

3. neither   4. Possible answer: $x = \dfrac{2x + 3}{x + 2}$

5. $x^2y = 20$ 6. $bh = 36$ 7. I, III

8. $C(x) = \dfrac{8}{x + 20}$ 9. y-axis   10. $x = -3, 1$

11. $x = 0, 1, 2$

12.
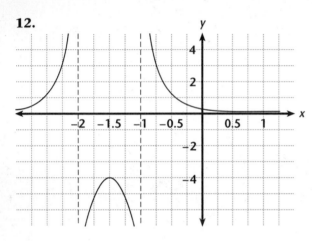

13.
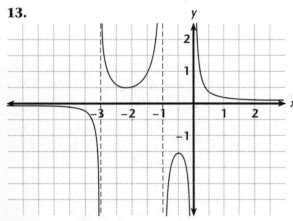

14. $x = -3$ 15. $\dfrac{x + 4}{x - 1}$ 16. $\dfrac{6x^3 + 3x^2 - 25}{x^2(x^3 - 5)}$

17. $x = 0.5, 1$ 18. $x = -\dfrac{9}{7}$

Alternative Assessment — Chapter 9

Form A

1. The graph of $y = \dfrac{1}{x}$ is translated 2 units to the left in $f(x) = \dfrac{1}{x + 2}$. Therefore, the horizontal asymptote is $y = 0$, and the vertical asymptote is $x = -2$.

2.
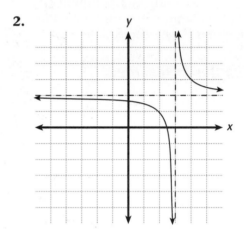

The graphs are equivalent. The domain and range for f are the same as the domain and range of g. To graph f, translate $y = \dfrac{1}{x}$ horizontally 3 units to the right and vertically 2 units up. Thus, the vertical asymptote is $x = 3$ and the horizontal asymptote is $y = 2$.

3.
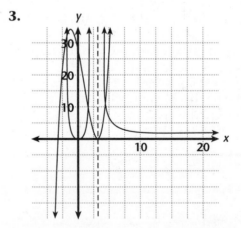

4. The zeros of p are $x = \pm 3$. The zeros of p are the vertical asymptotes of q. The domain of q is all real numbers $x \neq \pm 3$.

ANSWERS

5. The turning points of p are $x = -1$ and $x = 3$. A maximum of f occurs at $x = -1$. A minimum of q occurs at $x = -1$. A minimum of f occurs at $x = 3$. $x = 3$ is a vertical asymptote of q. Therefore, q has no maximum.

Score Point 4: Distinguished

The student demonstrates a comprehensive understanding of reciprocals of polynomial functions. The student uses perceptive, creative, and complex mathematical reasoning throughout the task. He or she is able to use sophisticated, precise, and appropriate mathematical language throughout the task. Theoretical knowledge is apparent and applied to concrete situations as the student successfully demonstrates a comprehensive understanding of core concepts throughout the task.

Score Point 3: Proficient

The student demonstrates a broad understanding of reciprocals of polynomial functions. The student uses perceptive mathematical reasoning most of the time. He or she is able to use precise and appropriate mathematical language most of the time. Theoretical knowledge is apparent and applied to concrete situations as the student attempts to draw conclusions based on his or her investigations.

Score Point 2: Apprentice

The student demonstrates an understanding of reciprocals of polynomial functions. He or she uses mathematical reasoning at times during the task. He or she uses appropriate mathematical language some of the time. Student attempts to apply theoretical knowledge to the task but may be able to draw conclusions from his or her investigation.

Score Point 1: Novice

The student demonstrates a basic understanding of reciprocals of polynomial functions. He or she uses appropriate mathematical language some of the time. Theoretical knowledge is extremely weak and many responses are irrelevant or illogical. He or she may fail to follow directions and has great difficulty in communicating his or her responses.

Score Point 0: Unsatisfactory

Student fails to make an attempt to complete the task and his or her responses are just an attempt to fill the page or restate the problem.

Form B

1. Graph $y = \frac{x - 3}{x}$ and $y = \frac{x - 4}{x - 2}$. Use trace to find the point of intersection. The solution is $x = 6$.

2. Convert the rational equation into a polynomial equation and find the roots. Extraneous roots are possible. Therefore, all roots must be checked. The solutions are $x = \frac{1}{3}$ or $x = \frac{-1}{2}$.

3. The graphing method shows no solution. The algebraic method yields the extraneous root $x = 1$. There is no solution to the rational equation.

4. The domain of f is all real numbers except $x = 0$. The domain of g is all real numbers except $x = 0$.

5. $h(x) = 1, x \neq 0$. When you simplify the sum of two rational functions, the domain of the sum function is still restricted by the domain of the two rational functions.

ANSWERS

Score Point 4: Distinguished

The student demonstrates a comprehensive understanding of solving rational functions. The student uses perceptive, creative, and complex mathematical reasoning throughout the task. He or she is able to use sophisticated precise, and appropriate mathematical language throughout the task. Theoretical knowledge is apparent and applied to concrete situations as the student successfully demonstrates a comprehensive understanding of core concepts throughout the task.

Score Point 3: Proficient

The student demonstrates a broad understanding of solving rational functions. The student uses perceptive mathematical reasoning most of the time. He or she is able to use precise and appropriate mathematical language most of the time. Theoretical knowledge is apparent and applied to concrete situations as the student attempts to draw conclusions based on his or her investigations.

Score Point 2: Apprentice

The student demonstrates an understanding of solving rational functions. He or she uses mathematical reasoning at times during the task. He or she uses appropriate mathematical language some of the time. Student attempts to apply theoretical knowledge to the task but may be able to draw conclusions from his or her investigation.

Score Point 1: Novice

The student demonstrates a basic understanding of solving rational functions. He or she uses appropriate mathematical language some of the time. Theoretical knowledge is extremely weak and many responses are irrelevant or illogical. He or she may fail to follow directions and has great difficulty in communicating his or her responses.

Score Point 0: Unsatisfactory

Student fails to make an attempt to complete the task and his or her responses are just an attempt to fill the page or restate the problem.

Practice & Apply—Chapter 10

Lesson 10.1

1. $y = \frac{-x^2}{16}$ **2.** $x = \frac{y^2}{16}$

3. vertex: $(-2, -1)$; axis: $x = -2$; focus $\left(-2, -\frac{11}{12}\right)$; directrix: $y = -\frac{13}{12}$

4. vertex: $(3, 1)$; axis: $y = 1$; focus: $\left(\frac{25}{8}, 1\right)$; directrix $x : x = \frac{23}{8}$

5.

Grade Stake Distance	0	50	100	150	200
Offset	0	0.625	2.5	5.625	10

6.

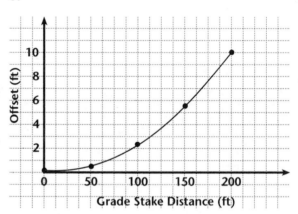

Lesson 10.2

1. $x^2 + y^2 = 25$ **2.** $(x - 5)^2 + y^2 = 25$

3. $(x + 5)^2 + (y - 2)^2 = 25$

4. $x^2 + (y - 6)^2 = 25$

5. Center: $(0, 0)$; radius: 2

6. Center: $(-1, -5)$; radius: 2

7. Center: $(0, -5)$; radius 2

8. Center: $(0, 5)$; radius 2

9. $(x - 1)^2 + y^2 = 25$

ANSWERS

- -

Lesson 10.3

1. $\dfrac{x^2}{9} + \dfrac{y^2}{16} = 1$ **2.** $\dfrac{(x-3)^2}{16} + \dfrac{(y-4)^2}{9} = 1$

3. $\dfrac{(x+3)^2}{9} + \dfrac{y^2}{16} = 1$

4. Vertices: $(-9, 0)$, $(3, 0)$; Co-vertices $(-3, 3)$, $(-3, -3)$

5. Vertices: $(1, 9)$, $(1, -3)$; Co-vertices $(-2, 3)$, $(4, 3)$

6. $(4, 0)$, $(-4, 0)$ **7.** $(0, 8)$, $(0, -8)$

8. $\dfrac{x^2}{22350.25} + \dfrac{y^2}{22344} = 1$ in millions of kilometers

9. The distance is 5 million kilometers.

10. $\dfrac{4x^2}{25} + \dfrac{y^2}{4} = 1$ where $-2.5 \le x \le 2.5$ and $0 \le y \le 2$

Lesson 10.4

1. $\dfrac{(x+1)^2}{16} - \dfrac{(y-3)^2}{9} = 1$ **2.** $\dfrac{y^2}{16} - x^2 = 1$

3. Center: $(1, 4)$; transverse axis: $(-1, 4)$, $(3, 4)$; conjugate axis: $(1, 9)$, $(1, -1)$; foci: $(1 \pm \sqrt{29}, 4)$

4. Center: $(0, 0)$; transverse axis: $(0, \pm 1)$; conjugate axis: $(\pm 2, 0)$; foci: $(0, \pm \sqrt{5})$

5. $\dfrac{y^2}{9} - \dfrac{x^2}{16} = 1$ **6.** $(y+6)^2 - (x-2)^2 = 1$

7. $\dfrac{x^2}{4} - \dfrac{y^2}{32} = 1$

8. $(y+6)^2 - (x-2)^2 = 1$; center: $(2, -6)$

9. $\dfrac{x^2}{2500} - \dfrac{y^2}{20000} = 1$

Lesson 10.5

1. $(0, 3)$, $(4, 0)$ **2.** $(3, 2)$, $(1.6, 3.7)$

3. $(2, 0)$, $(4, -12)$ **4.** $(4, 0)$, $(0, 4)$

5. $(-4, 0)$, $(5, 0.75)$

6. $(2, 2)$, $(2, -2)$, $(-2, 2)$, $(-2, -2)$

7. $(0, 4)$ **8.** $(-3, 0)$ **9.** 50 ft by 75 ft

10. 5 units or 20 units

Lesson 10.6

1. $x(t) = 5\cos t$
$y(t) = 5\sin t$

2. $x(t) = 5 \cos t$
$y(t) = 4 \sin t$

3. Answers may vary. Possible answer:
$x(t) = 2t$
$y(t) = 16t^2 - 2t$

4. $\dfrac{x^2}{16} + \dfrac{y^2}{25} = 1$; ellipse

5. $x = y - 2y^2$; parabola

6. $x^2 + y^2 = 16$; circle

7. $(x-3)^2 + (y-5)^2 = 16$; circle, center $(3, 5)$, radius 4

8. $(x+3)^2 + \dfrac{(y-5)^2}{16} = 1$; ellipse, center $(-3, 5)$, vertices $(-3, 9)$, $(-3, 1)$, $(-4, 5)$, $(-2, 5)$

9. $x = 5\cos t$
$y = 3 + 5\sin t$

10. $x = 5 \cos t$
$y = -3 + \sin t$

11. $x = 3 \cos t$
$y = 5 \sin t$

12. $x = 2 \cos t$
$y = 6 \sin t$

Enrichment — Chapter 10

Lesson 10.1

1. The vertex of its parabolic path.

2. 4.5 ft **3.** 6 in. from the vertex **4.** 27 ft.

ANSWERS

5. Student responses should include substituting $t = \frac{x}{40}$ into the equation for y, giving $y = \frac{-x^2}{100}$, which is the equation of a parabola.

6. 24.5 m^2 7. 25 8. $1000 9. 10

Lesson 10.2

1. $x^2 + y^2 = 6,250,000$

2. Possible answer: The part that includes the shore area.

3. $(x - 1000)^2 + y^2 = 9,000,000$

4.

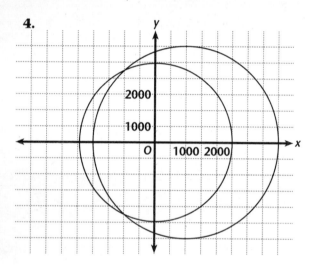

5. Student responses should indicate that only one of the locations is the water, the other is on the shore.

6. 3000 cm 7. $x^2 + y^2 = 9$

8. $x^2 + (y - 4)^2 = 25$

Lesson 10.3

1. average distance = half the length of the major axis.

2. about 1.84 y 3. ≈ 2.17 AU

4. $\dfrac{x^2}{21,160,000} + \dfrac{y^2}{21,000,000} = 1$

5. 12 ft 6. $\dfrac{8}{3}$

Lesson 10.4

1. $\dfrac{x^2}{4,000,000} - \dfrac{y^2}{21,000,000} = 1$

2.

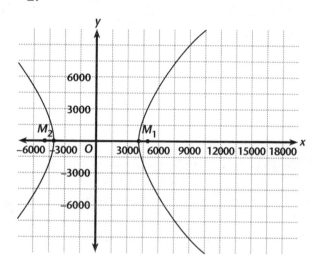

3. Student responses should indicate that the intersection of the two graphs will narrow down the location.

4.

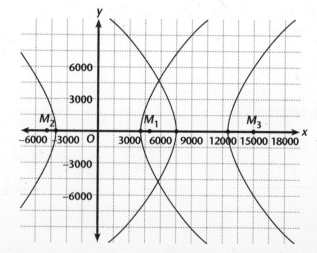

ANSWERS

5. Possible answers: The time differences for M_2 and M_3; data from another microphone.

Lesson 10.5

1. no points, one point, two points

2. no points, one point, two points

3. no points, one point, two points, three points, four points

4. $(-4, -3)$, $(-4, 3)$, $(4, -3)$ $(4, 3)$

5. 3 m by 18 m **6.** $\sqrt{5}$ m, 2 m

Lesson 10.6

1. hyperbola **2.** ellipse **3.** ellipse

4. ellipse **5.** hyperbola **6.** ellipse

7. ellipse **8.** parabola **9.** ellipse

10. ellipse **11.** ellipse **12.** parabola

13. hyperbola **14.** ellipse **15.** ellipse

16. ellipse **17.** hyperbola **18.** hyperbola

Technology — Chapter 10

Lesson 10.1

1.

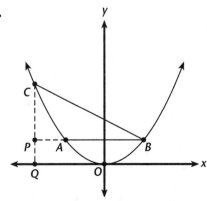

If C is above $\overline{AB}$, then $CP = CQ - PQ = x^2 - 4$.

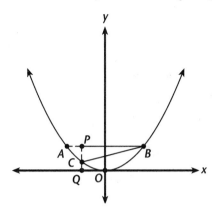

If C is below $\overline{AB}$, then $CP = PQ - CQ = 4 - x^2$. So, $CP = |x^2 - 4|$.

2.

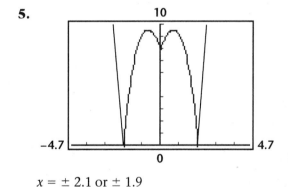

$x = \pm 2.2$ or ± 1.8

3. four, three, two **4.** $K = |x^2 - 4|(|x| + 2)$

5.

$x = \pm 2.1$ or ± 1.9

Lesson 10.2

1. $y = \sqrt{9 - x^2}$ or $y = -\sqrt{9 - x^2}$

2. $d = \sqrt{(x + 1)^2 + (\sqrt{9 - x^2} - 1)^2}$ or

$d = \sqrt{(x + 1)^2 + (-\sqrt{9 - x^2} - 1)^2}$

3. $-3 \le x \le 3$ **4.** $x \approx -2.1; y \approx -2.1$

5. $x \approx 2.1; y \approx 2.1$

Lesson 10.3

1. Area $= 4|x| \sqrt{\dfrac{144 - 9x^2}{16}}$

2.

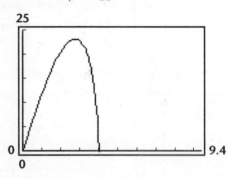

3. $x \approx 2.8$; Area ≈ 24

4. Area $= 4|x| \sqrt{\dfrac{100 - 4x^2}{25}}$

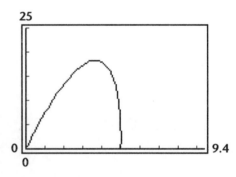

5. $x \approx 3.5$; Area ≈ 20

6. If $ax^2 + by^2 = ab$, the maximum area of $ABCD$ seems to be $2\sqrt{ab}$.

Lesson 10.4

1–4.

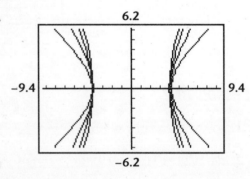

5. Each graph is a hyperbola with vertices $(-4, 0)$ and $(4, 0)$. As c increases the branches become more shallow. In other words, they approach the vertical lines $x = -4$ and $x = 4$.

6–9.

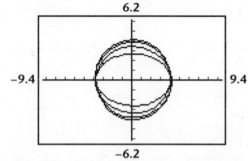

10. When $-4 < c < 4$, the graph is no longer a hyperbola, but rather an ellipse. When $c = 0$, the ellipse becomes a circle centered at the origin and radius 4.

Lesson 10.5

1. $\begin{bmatrix} 0.1053 & 0.2632 \\ 0.1579 & -0.1053 \end{bmatrix} \begin{bmatrix} 2.6316 \\ 0.9474 \end{bmatrix}$
$(\pm 1.6222, \pm 0.9733)$

2. $(\pm 3.3072, \pm 2.2500)$ **3.** $(\pm 4, 0)$

4. no solution **5.** $(\pm 4, 0)$ **6.** no solution

7. $(\pm 2.9155, \pm 2.7386)$

8.

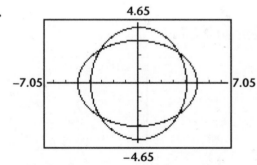

Yes, the graph confirms the results in Exercise 2.

ANSWERS

9.

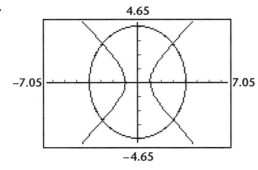

Yes, the graph confirms the results in Exercise 7.

Lesson 10.6

1.

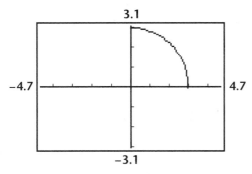

2.

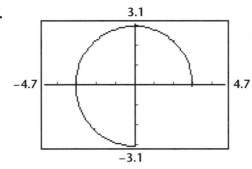

3. $0 \le t \le 2\pi$

4.

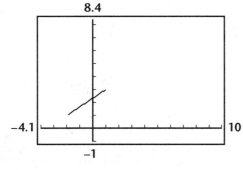

5.

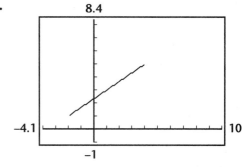

6.

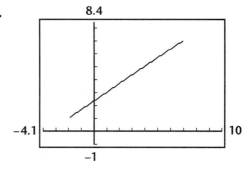

7. You will get a line segment whose endpoints are $(x(a), y(a))$ and $(x(b), y(b))$.

8.

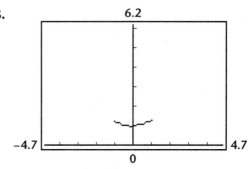

9.

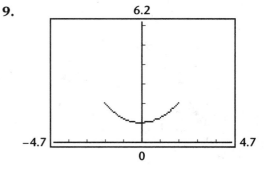

ANSWERS

10.

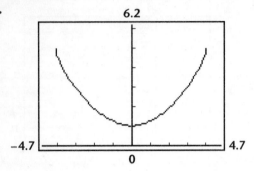

11. You will get a segment of a parabola, a finite portion of the parabola whose endpoints are $(x(-a), y(-a))$ and $(x(a), y(a))$.

Lesson Activities — Chapter 10

Lesson 10.1

1. Vertex is the point midway between point *F* and the initial line. Focus is point *F*. Directrix is the initial line.

2. $y = ax^2$

3.

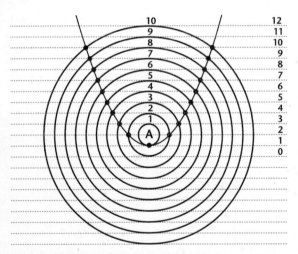

4. The curve is the set of points in a plane that are the same distance from a fixed line.

5. The vertex is on the axis of symmetry, one unit from point *A* and intersecting line 1. The focus is point *A*. The directrix is line 0.

6. Check students' graphs.

Lesson 10.2

1. Cycloid.

2.

t	*x*	*y*
0	0	0
0.5π	0.57	1
π	3.14	2
1.5π	5.71	1
3π	9.42	2
4π	12.57	0

3. four cycles   **4.** 0 to 16π

Lesson 10.3

1.

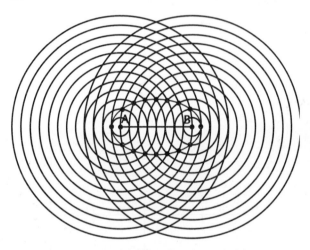

2. The curve is a set of points in a plane the sum of whose distances from two fixed points is constant.

3. The center is the midpoint of line segment *AB*. The foci are point *A* and point *B*. The vertices are on the intersection of circles 9 and 1.

4. $\dfrac{x^2}{25} + \dfrac{y^2}{9} = 1$

ANSWERS

5. Possible answer:

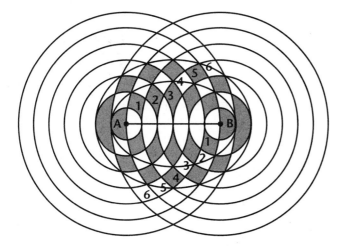

6. an ellipse

Lesson 10.4

1.

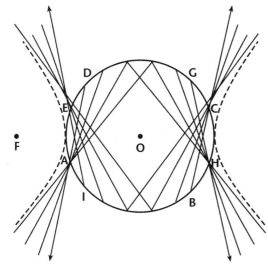

2. The curve is a set of points in a plane such that the difference of the distances from two fixed points to any point on the hyperbola is constant.

3. 2 units (the length of the diameter)

4.

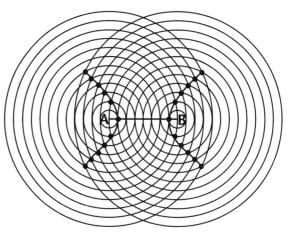

5. Center: the midpoint of line segment *AB*; foci: point *A* and point *B*.

6. $\dfrac{x^2}{9} - \dfrac{y^2}{25} = 1$

Lesson 10.5

1. square (or diamond)   **2.** octagon

3. decagon   **4.** $\left(\dfrac{2}{3}\right)\pi$ **5.** pentagon

6. $\left(\dfrac{2}{7}\right)\pi$ will produce a septagon; $\left(\dfrac{2}{9}\right)\pi$ will produce a nonagon.

7. 0 to 4π **8.** 0 to 8π

Lesson 10.6

1.

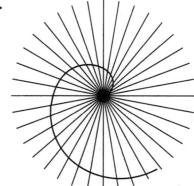

2. Possible answer: The graph spirals out from the center in a clockwise direction.

3. Possible answer: The spiral lines are tighter or closer together.

ANSWERS

4. Possible answer: As the coefficient decreases the "width" (diameter) of the spiral decreases; more points are plotted.

5. The spiral is graphed counterclockwise.

6. T represents the radius of each circle formed by the spiral.

Assessment — Chapter 10

Assessing Prior Knowledge 10.1

1. $y = \frac{3}{10}$ 2. $y = 5$ 3. $(y + 5)^2$

Quiz 10.1

1. $y = -\frac{1}{14}(x^2 - 7)$ 2. $x = \frac{1}{8}y^2$

3. Vertex: $(0, 0)$; Focus: $(1, 0)$;
   Directrix: $x = -1$; opens upward

4. Vertex: $(0, 0)$; Focus: $\left(0, \frac{5}{2}\right)$;
   Directrix: $y = -\frac{5}{2}$; opens to the right

5. Vertex: $\left(-1, \frac{5}{8}\right)$; Focus: $\left(-1, \frac{21}{8}\right)$;
   Directrix: $y = -\frac{11}{8}$; opens upward

6. $V = (25, 40)$; $Y = \frac{-8}{125}(x - 25)^2 + 40$

Assessing Prior Knowledge 10.2

1. $(y + 6)^2$ 2. 3.6 units

3. $\sqrt{(y - 2)^2 + (x - 1)^2}$

Quiz 10.2

1. $(x - 2)^2 + (y - 4)^2 = 25$

2. $(x + 6)^2 + (y - 4)^2 = \frac{25}{4}$

3. $\left(x - \frac{3}{5}\right)^2 + \left(y + \frac{1}{3}\right)^2 = 5$

4. $(x + 4)^2 + (y + 3)^2 = 16$

5. $C: (-2, -1); r: 1$ 6. $C: (2, 4); r: 2\sqrt{6}$

7. $C: (-2, 1); r: \sqrt{11}$ 8. $C: (-3, 4); r: 4\sqrt{2}$

9. $(x - 6)^2 + (y - 1)^2 = 4$

Assessing Prior Knowledge 10.3

1. $1, -5$ 2. $-1, 3$ 3. $1, -7$ 4. $2 \pm i\sqrt{2}$

Quiz 10.3

1. $\frac{x^2}{9} + \frac{y^2}{25} = 1$ 2. $\frac{x^2}{64} + \frac{y^2}{16} = 1$

3. $\frac{9x^2}{16} + \frac{4y^2}{81} = 1$

4. $V: \left(2 \pm \frac{\sqrt{26}}{2}, -\frac{3}{2}\right)$; Co-V: $\left(2, -\frac{3}{2} \pm \frac{\sqrt{13}}{2}\right)$

5. $V: \left(\frac{5}{2}, \frac{-3 \pm 3\sqrt{6}}{2}\right)$; Co-V: $\left(\frac{5}{2} \pm \frac{3\sqrt{30}}{10}, \frac{-3}{2}\right)$

6. $\frac{x}{(11,500)^2} + \frac{y}{(9487)^2} = 1$

Assessing Prior Knowledge 10.4

1. $\frac{x^2}{16} + \frac{y^2}{9} = 1$ 2. $\frac{x^2}{81} + \frac{y^2}{49} = 1$

Quiz 10.4

1. $\frac{x^2}{25} - \frac{y^2}{9} = 1$ 2. $\frac{y^2}{64} - \frac{x^2}{11} = 1$

3. $\frac{(x - 1)^2}{4} - \frac{y^2}{12} = 1$

4. $\frac{(x - 3)^2}{4} - \frac{(y - 3)^2}{12} = 1$

5. $\frac{(x + 5)^2}{9} - \frac{(y - 3)^2}{9} = 1$; $C: (-5, 3)$;
   transverse axis: 6; conjugate axis: 6;
   Foci: $(-5 + 3\sqrt{2}, 3), (-5 - 3\sqrt{2}, 3)$

Mid-Chapter Assessment

1. a   2. b   3. b   4. a

ANSWERS

---

5. $y = -\frac{2}{9}(x - 15)^2 + 50$

6. $(x + 3)^2 + (y + 5)^2 = 16$

7. V: $(-6, 3), (2, 3)$; Co-V: $(-2, 6), (-2, 0)$

8. V: $(4, 3)$; F: $\left(4, 3\frac{1}{2}\right)$

9. C: $(3, -2)$; V: $(3, 0)$ and $(3, -4)$;
F: $(3 \pm \sqrt{7}, -2)$

Assessing Prior Knowledge 10.5

1. $a = 5$ **2.** $a = \frac{1}{4}$ **3.** $a = -9$

Quiz 10.5

1. $(3, -2); \left(\frac{1}{5}, \frac{18}{5}\right)$ **2.** $(4, 12); (-1, -3)$

3. $(2, 2), (2, -2), (-2, 2), (-2, -2)$

4. $(0, 4), (0, -4)$

5. $(2, 5), (5, 2), (-2, -5), (-5, -2)$

6. $(\sqrt{3}, \sqrt{6}), (\sqrt{3}, -\sqrt{6}), (-\sqrt{3}, \sqrt{6}),$
$(-\sqrt{3}, -\sqrt{6})$

7. 14 in. $\times$ 16 in.

Assessing Prior Knowledge 10.6

1. $\sec^2 \theta$ **2.** $\cos^2 \theta$ **3.** $\cos \theta$

Quiz 10.6

1. $\begin{cases} x(t) = 4 \cos t \\ y(t) = 4 \sin t \\ \text{circle} \end{cases}$

2. $\begin{cases} x(t) = 4\cos t \\ y(t) = 5\sin t \\ \text{ellipse} \end{cases}$

3. $\begin{cases} x(t) = t \\ y(t) = t^2 + 5 \\ \text{parabola} \end{cases}$

4. $\begin{cases} x(t) = \cos t + 3 \\ y(t) = \sin t + 2 \\ \text{circle} \end{cases}$

5. $\begin{cases} x(t) = 5 \cos t + 2 \\ y(t) = 4 \sin t - 3 \\ \text{ellipse} \end{cases}$

6. $\frac{x^2}{49} + \frac{y^2}{9} = 1$; ellipse   **7.** $x^2 + y = 64$; circle

Chapter Assessment, Form A

1. d   **2.** b   **3.** c   **4.** c   **5.** c   **6.** a   **7.** b

8. b   **9.** d   **10.** c   **11.** d   **12.** a   **13.** b

14. b   **15.** a   **16.** d   **17.** c   **18.** a

Chapter Assessment, Form B

1. $y = \frac{1}{8}x^2 - 1$ **2.** $(2, -1)$

3. $y = \frac{1}{8}(x - 2)^2 + 3$

4. $(x + 3)^2 + (y - 4)^2 = 25$

5. Center: $(4, -3)$; Radius: 6

6. $(x + 3)^2 + (y + 1)^2 = 16$ **7.** $\frac{x^2}{16} + \frac{y^2}{64} = 1$

8. $(0, \pm 5), (\pm 2, 0)$ **9.** $\frac{x^2}{5} + \frac{y^2}{9} = 1$

10. 18 cm $\times$ 24 cm   **11.** $\frac{x^2}{16} - \frac{y^2}{64} = 1$

12. $\frac{y^2}{3} - \frac{x^2}{75} = 1$ **13.** $\left(0, \pm \frac{5}{4}\right)$

14. $(3, 4), (-4, -3)$ **15.** $(-1, 1), \left(\frac{1}{4}, \frac{9}{4}\right)$

16. $\begin{cases} x(t) = 9 \cos t \\ y(t) = 9 \sin t \end{cases}$

17. $(x + 2)^2 + (y + 3)^2 = 16$ **18.** circle

ANSWERS

Alternative Assessment — Chapter 10

Form A

1. Write the equation in standard form by completing the square. In standard form the equation of the parabola is $y = \frac{1}{8}(x - 3)^2 + 2$. The parabola opens upward because $a = \frac{1}{8}$ is greater than zero. The vertex is $(3, 2)$, the focus is $(3, 4)$ and the directrix is $y = 0$.

2. Write the equation of the circle in standard form by completing the square. In standard form the equation of the circle is $(x - 3)^2 + (y - 1)^2 = 4$. The center is $(3, -1)$ and the radius is 2.

3. The equation in standard form is $\frac{x^2}{12} + \frac{y^2}{16} = 1$. The center is $(0, 0)$. The vertices are $(0, 4)$ and $(0, -4)$. The foci are $(0, -2)$ and $(0, 2)$. From the standard equation of the ellipse, the longer axis is the vertical axis.

4. $\frac{x^2}{25} + \frac{y^2}{9} = 1$

Score Point 4: Distinguished

The student demonstrates a comprehensive understanding of equations representing a parabola, a circle, and an ellipse. The student uses perceptive, creative, and complex mathematical reasoning throughout the task. He or she is able to use sophisticated, precise, and appropriate mathematical language throughout the task. Theoretical knowledge is apparent and applied to concrete situations as the student successfully demonstrates a comprehensive understanding of core concepts throughout the task.

Score Point 3: Proficient

The student demonstrates a broad understanding of equations representing a parabola, a circle, and an ellipse. The student uses perceptive mathematical reasoning most of the time. He or she is able to use precise and appropriate mathematical language most of the time. Theoretical knowledge is apparent and applied to concrete situations as the student attempts to draw conclusions based on his or her investigations.

Score Point 2: Apprentice

The student demonstrates an understanding of equations representing a parabola, a circle, and an ellipse. He or she uses mathematical reasoning at times during the task. He or she uses appropriate mathematical language some of the time. Student attempts to apply theoretical knowledge to the task but may be able to draw conclusions from his or her investigation.

Score Point 1: Novice

The student demonstrates a basic understanding of equations representing a parabola, a circle, and an ellipse. He or she uses appropriate mathematical language some of the time. Theoretical knowledge is extremely weak and many responses are irrelevant or illogical. He or she may fail to follow directions and has great difficulty in communicating his or her responses.

Score Point 0: Unsatisfactory

Student fails to make an attempt to complete the task and his or her responses are just an attempt to fill the page or restate the problem.

Form B

1. A circle and an ellipse can intersect in 0, 1, 2, 3, or 4 points.

2. The graphs of a first-degree equation and a second-degree equation can intersect in 0, 1, or 2 points.

ANSWERS

---

3. Graph $y_1 = \sqrt{(16 + x^2)}$, $y_2 = -\sqrt{(16 + x^2)}$, $y_3 = \sqrt{(34 - x^2)}$, and $y_4 = -\sqrt{(34 - x^2)}$. The solutions are $(3, 5)$, $(3, -5)$, $(-3, 5)$, and $(-3, -5)$.

4. The solutions are $(2\sqrt{2}, \sqrt{2})$ and $(-2\sqrt{2}, -\sqrt{2})$.

Score Point 4: Distinguished

The student demonstrates a comprehensive understanding of solving non-linear systems of equations. The student uses perceptive, creative, and complex mathematical reasoning throughout the task. He or she is able to use sophisticated, precise, and appropriate mathematical language throughout the task. Theoretical knowledge is apparent and applied to concrete situations as the student successfully demonstrates a comprehensive understanding of core concepts throughout the task.

Score Point 3: Proficient

The student demonstrates a broad understanding of solving non-linear systems of equations. The student uses perceptive mathematical reasoning most of the time. He or she is able to use precise and appropriate mathematical language most of the time. Theoretical knowledge is apparent and applied to concrete situations as the student attempts to draw conclusions based on his or her investigations.

Score Point 2: Apprentice

The student demonstrates an understanding of solving non-linear systems of equations. He or she uses mathematical reasoning at times during the task. He or she uses appropriate mathematical language some of the time. Student attempts to apply theoretical knowledge to the task but may be able to draw conclusions from his or her investigation.

Score Point 1: Novice

The student demonstrates a basic understanding of solving non-linear systems of equations. He or she uses appropriate mathematical language some of the time. Theoretical knowledge is extremely weak and many responses are irrelevant or illogical. He or she may fail to follow directions and has great difficulty in communicating his or her responses.

Score Point 0: Unsatisfactory

Student fails to make an attempt to complete the task and his or her responses are just an attempt to fill the page or restate the problem.

Cumulative Assessment Free-Response Grids

Exercise _____

Exercise _____

Exercise _____

Exercise _____

Exercise _____

Exercise _____

Exercise _____

Exercise _____

Exercise _____

HRW Advanced Algebra

Free Response Grid

PORTFOLIO HOLISTIC SCORING GUIDE

An individual portfolio is likely to be characterized by some, but not all, of the descriptors for a particular level. Therefore, the overall score should be the level at which the appropriate descriptors for a portfolio are clustered.

		NOVICE	APPRENTICE	PROFICIENT	DISTINGUISHED
PROBLEM SOLVING	Understanding/Strategies	• Indicates a basic understanding of problems and uses strategies	• Indicates an understanding of problems and selects appropriate strategies	• Indicates a broad understanding of problems with alternate strategies	• Indicates a comprehensive understanding of problems with efficient, sophisticated strategies
	Execution/Extensions	• Implements strategies with minor mathematical errors in the solution without observations or extensions	• Accurately implements strategies with solutions, with limited observations or extension	• Accurately and efficiently implements and analyzes strategies with correct solutions, with extension	• Accurately and efficiently implements and evaluates sophisticated strategies with correct solutions and includes analysis, justifications, and extensions
REASONING		• Uses mathematical reasoning	• Uses appropriate mathematical reasoning	• Uses perceptive mathematical reasoning	• Uses perceptive, creative, and complex mathematical reasoning
MATHEMATICAL COMMUNICATION	Language	• Uses appropriate mathematical language some of the time	• Uses appropriate mathematical language	• Uses precise and appropriate mathematical language most of the time	• Uses sophisticated, precise, and appropriate mathematical language throughout
	Representations	• Uses few mathematical representations	• Uses a variety of mathematical representations accurately and appropriately	• Uses a wide variety of mathematical representations accurately and appropriately; uses multiple representations within some entries	• Uses a wide variety of mathematical representations accurately and appropriately; uses multiple representations within entries and states their connections
UNDERSTANDING/CONNECTING CORE CONCEPTS		• Indicates a basic understanding of core concepts	• Indicates an understanding of core concepts with limited connections	• Indicates a broad understanding of some core concepts with connections	• Indicates a comprehensive understanding of core concepts with connections throughout
TYPES AND TOOLS		• Includes few types; uses few tools	• Includes a variety of types; uses tools appropriately	• Includes a wide variety of types; uses a wide variety of tools appropriately	• Includes all types; uses a wide variety of tools appropriately and insightfully

PERFORMANCE DESCRIPTORS

PROBLEM SOLVING
- Understanding the features of a problem (understands the question, restates the problem in own words)
- Explores (draws a diagram, constructs a model and/or chart, records data, looks for patterns)
- Selects an appropriate strategy (guesses and checks, makes an exhaustive list, solves a simpler but similar problem, works backward, estimates a solution)
- Solves (implements a strategy with an accurate solution)
- Reviews, revises, and extends (verifies, explores, analyzes, evaluates strategies/solutions; formulates a rule)

REASONING
- Observes data, records and recognizes patterns, makes mathematical conjectures (inductive reason)
- Validates mathematical conjectures through logical arguments or counter-examples; constructs valid arguments (deductive reasoning)

MATHEMATICAL COMMUNICATION
- Provides quality explanations and expresses concepts, ideas, and reflections clearly
- Uses appropriate mathematical notation and terminology
- Provides various mathematical representations (models, graphs, charts, diagrams, words, pictures, numerals, symbols, equations)

UNDERSTANDING/CONNECTING CORE CONCEPTS
- Demonstrates an understanding of core concepts
- Recognizes, makes, or applies the connections among the mathematical core concepts to other disciplines, and to the real world

WORKSPACE/ANNOTATIONS

PORTFOLIO CONTENTS
- Table of Contents
- Letter to Reviewer
- 5–7 Best Entries

BREADTH OF ENTRIES

TYPES
- INVESTIGATIONS/DISCOVERY
- APPLICATIONS
- NON-ROUTINE PROBLEMS
- PROJECTS
- INTERDISCIPLINARY
- WRITING

TOOLS
- CALCULATORS
- COMPUTER AND OTHER TECHNOLOGY
- MODELS-MANIPULATIVE
- MEASUREMENT INSTRUMENTS
- OTHERS

GROUP ENTRY

Place an X on each continuum to indicate the degree of understanding demonstrated for each core concept.

DEGREE OF UNDERSTANDING OF CORE CONCEPTS

	Basic				Comprehensive with connections
NUMBER					
MATHEMATICAL PROCEDURES					
SPACE & DIMENSIONALITY					
MEASUREMENT					
CHANGE					
MATHEMATICAL STRUCTURE					
DATA: STATISTICS AND PROBABILITY					

The Kentucky Mathematics Portfolio was developed by the Kentucky Department of Education for use by school districts throughout that state.